New Directions in Biology Teaching

perspectives for the 1980s

edited by
faith m. hickman
jane butler kahle

national association of biology teachers
1982

Library of Congress Cataloging in Publication Data

Main entry under title:

New directions in biology teaching.

Includes bibliographies.
1. Biology—Study and teaching. I. Hickman, Faith, M., 1946– . II. Kahle, Jane Butler.
QH315.N43 574'.07'1 81-16788
ISBN 0-941313-02-5 AACR2
ISBN 0-941212-01-7 (pbk.)

Printed in the United States of America

10 9 8 7 6 5 4 3 2 1

(Typeset and printed by Interstate Printers & Publishers, Inc., Danville, IL.)

New Directions in Biology Teaching

CONTENTS

FOREWORD

The general high school biology course is about eighty years old. Emerging as a replacement for botany, human physiology, and zoology, and at first strongly reflecting this tripartite origin, it has become nearly encyclopedic in scope as each decade added more knowledge. Originally taken by only a few percent of high school students, many of whom did not go on to college, it has become the one science course taken by most high school graduates.

Despite an appearance of stability, the content, organization, and emphasis of high school biology has changed through the years in response to diverse social forces. Paul DeHart Hurd noted in the *Biological Sciences Curriculum Study Bulletin* No. 1 (1961) that during the period 1910-1920 emphasis was placed on human welfare, including, ". . . vocations, health, sanitation, avocations, appreciations and understanding the environment." With the twenties, standardized tests became popular, leading to pressure for a "standard" biology course with emphasis on the principles of biology. The Great Depression decade focused on meeting student needs, with health, consumer education, and social welfare emerging as organizing ideas for biology curricula. Learning theory became a factor in curriculum development. The years surrounding World War II radically changed our perception of power and potential for growth, while in biology education conflict arose between efforts to prepare students for careers in science while providing a "useful" biology course for the majority of students.

Hurd describes the decade from 1950 to 1960 as one of confusion and crisis in biology education. Conflicting pressures came from the needs of national defense, industry, and higher education, while society at large produced a booming teenage population, belief in college education, and fascination with the coming space age. The three biology curricula of BSCS along with laboratory blocks and supporting materials, were in trial use by the end of this decade; they emphasized new knowledge, research methods, and application of learning theory. The

BSCS materials came to dominate biology education in the sixties, leading to revisions of traditional texts, incorporating the ideas pioneered with National Science Foundation support.

Hurd's survey of biology education ended with the sixties, but those reading the present volume in the eighties will have their own impression of the intervening decade. Development of an environmental ethic, concern for individual worth, and recognition that there are absolute limits to human population growth contrasted sharply with an underlying fear that we had somehow lost control. Science support was eroded by the retreat to basics, while new biological knowledge flooded the news. Schools became the focus of conflict between factions who wanted to implement *their* vision of society. Biology teachers, often demoralized, frequently called their own retreat to a version of "basic" biology, based on lecture, recitation, and textbooks adjusted to lower reading levels and expectations.

Which brings us to 1981 and the eighties. What forces are affecting biology education? What are appropriate responses for professional teachers? What biological knowledge should be captured by the 175-odd hours allotted to the average high school biology course? The authors contributing to this volume deal with these questions and their answers are of concern to all biology teachers.

NABT is proud to place *New Directions in Biology Teaching* next to *Social Implications of Biological Education,* edited by Arnold Grobman and published in 1970. Faith Hickman and Jane Kahle have done an outstanding job of editing a book which may well catalyze development in biological education for the next decade.

Wayne A. Moyer
Executive Director, NABT

Reference

HURD, P. D. 1961. *Biological education in American secondary schools 1890-1960.* Washington, D.C.: American Institute of Biological Sciences.

PREFACE

Life is a series of collisions with the future; it is not the sum of what we have been but what we yearn to be.

—José Ortega y Gasset

The task of preparing a volume with a theme as ambitious as "new directions in biology teaching" reminds one of the truth of this quote. The profession of teaching, like any other human endeavor, is indeed a series of collisions with the future. And the future is determined not by the past, but by our plans, expectations—yes, even our dreams—for tomorrow. Ward Madden said it best: "Perhaps the greatest impulse to trying to foresee and plan the future comes from the combination of having new tools with which to do it and the growing realization that every technological and social innovation has repercussions which spread like a wave through the complex interlocked sections of society." The teaching of biology is an important section that interlocks with many other elements of our world. Our interactions with members of our own generation and with new generations have ripples of influence that extend far beyond our own classrooms and our own time.

The papers that are included in this book have been prepared by individuals who seem to welcome our collisions with the future. The work represented here gives promise to the paths we may take and provides direction as we move toward the future that we can both envision and shape. Paul Hurd's "Biology for Life and Living" sets the context for this publication. Hurd sees in our future new demands, new problems, new expectations, and new challenges—the essence of new directions in biology teaching. Hurd, who is among the premiere historians and philosophers of science education, believes that a new emphasis on "life and living" is essential if students are to face the complexities of a future that will present to its inhabitants biological and social problems of an import scarcely imagined in the past.

Those problems present today's teachers with a critical question: What and how can I teach best to help prepare my students for their lives in the future? Hurd provides some answers, as do other contributors to this volume who address specific areas of educational concern. Kormondy looks at the importance of values and ethics in life science instruction. From differing perspectives, Glass and Omenn treat the interaction of education and science with matters of social and public policy. The relevance of biology for life and living is examined in two exemplar content areas, as Holtzman discusses the new human genetics and Roberts examines human sexuality.

The publication concludes with McInerney's treatment of the relationship of science and values; Hickman's look at the arguments and counterarguments concerning dealing with controversial issues in the classroom; and a summary by Bybee and Kahle of steps and priorities needed to move beyond the present teaching practices and emphases toward a new biology instruction for the future.

This volume has a special feature worthy of close attention. Each paper is followed by a series of commentaries prepared by high school and college teachers. These commentators were asked to address what they believed to be the significance of each paper in terms of its implications for the day-to-day practice of teaching. The commentaries are both theoretical and practical; and the commentators are not always in agreement either with the author of the paper or with each other. But the commentaries provide stimulus for further thought and exploration of new directions in biology teaching.

We have greatly enjoyed preparing this publication. We sincerely hope that teachers of biology and life science will find this book interesting and useful. We hope, also, that the community of life science education can join together in making the new directions outlined here a reality for today and tomorrow. We welcome your comments and suggestions.

The Editors

Faith M. Hickman
Biological Sciences Curriculum Study
Boulder, Colorado

Jane Butler Kahle
Purdue University
West Lafayette, Indiana

ACKNOWLEDGMENTS

The National Association of Biology Teachers and the editors of this special publication wish to thank the participating authors and the following members of the Commentary Committee for their valuable contributions. Special thanks also go to Cathrine M. Monson for her preparation and supervision of this publication and Eunice A. Combs for her editorial assistance. On the Reston staff, our thanks and commendation go to Diana Baber, who served as production editor, and Debora Bukovskey, who served as proofreader.

Jane Abbott
Waterville Senior High School
Waterville, Maine

Roy K. Baer
Wachusett Regional High School
Holden, Massachusetts

David Lapierre
Logansport High School
Logansport, Indiana

John G. Lepley
Fort Benton Schools
Fort Benton, Montana

Betty W. Long
Atherton High School
Louisville, Kentucky

Sister Marian Catherine McGrann
St. Bernard High School
Playa del Rey, California

James L. Mariner
Fountain Valley School
Colorado Springs, Colorado

James Meyer
El Camino Real High School
Woodland Hills, California

Gordon E. Uno
University of Oklahoma
Norman, Oklahoma

1

BIOLOGY FOR LIFE AND LIVING: PERSPECTIVES FOR THE 1980s

Paul DeHart Hurd

What, then, is the new biology for life and living all about? It is a biology that honors human nature and the character of human culture and that recognizes the cognitive potential of human beings.

The study of human beings has become the new challenge for biology teaching in the 1980s. The character of science is undergoing changes; science and technology have become the two sides of a single coin. We have witnessed the double marriage of science and technology to human welfare and to economic progress. The thrust today is to bring biological education into harmony with both science and society. The goal is to help students lead more meaningful, satisfying, and responsible lives.

Individually and collectively we have more information available to us than all generations who have lived before. Yet we seem to know less about how to solve contemporary problems of life and living. We should not continue to deny students the skills required to process and utilize knowledge for their own benefit and for the welfare of the social group.

We should also not overlook the fact that life is lived forward, not backward. A future-oriented biology course has meaning for interpreting problems and issues as they exist not only for today but also for tomorrow. Never have human beings been able to influence their future so decisively—yet, never have we been so uncertain.

The new context for biology grows out of human events, the character of the culture, and the concepts of biology encountered most widely in human experience. The curriculum must be a balance of science, culture, and humanity.

THE GREATEST DISCOVERY in the history of biology was made in the latter half of the twentieth century. A known but yet unknown animal species entered our perceptions. At the time of discovery that species was in some jeopardy of relatively early extinction. It may still be. Because we do not yet fully understand the adaptive capacity of this animal, its future is uncertain. Seldom has a new species aroused so much concern among biologists or stimulated so much curiosity in students. Who is this animal, so recently discovered, so greatly significant, so widely neglected? It is the species *Homo sapiens:* you and I—billions of us—and yet we are almost unknown.

Until a decade ago few biologists raised questions about humankind. Such questions were considered unscientific, especially if they were concerned with human potential and quality of life. Today, this unique, complex, social animal is our most widely investigated mammal. The study of human beings has become the new challenge for biology teaching in the 1980s.

But before a suitable curriculum can be developed there are questions that must be answered: What should a student learn about biology to live intelligently and effectively in the world of today and tomorrow? What experiences should a biology student have to develop a feeling of maturity, sound judgment, and effective action? What we need to know to answer these questions is not at all apparent, but considerable information is available. We do recognize, for example, that the subject matter of the curriculum should be specific to the culture and appropriate for the 1980s. Furthermore, recent changes in the nature of science as an enterprise, in biology as a discipline, and in society as a whole make it likely that the ideal biology program of a decade ago could be a travesty today.

The conditions that have brought about questioning of biology teaching are products of widespread cultural transformations and of changes in the character of science itself. Science and technology have become the two sides of a single coin. We have recently witnessed the double marriage of science and technology to human welfare and economic progress. The thrust of science into our everyday affairs, and its potential for increasing the quality of our life, has generated in the public mind a credibility gap between today's biology curriculum and the real world of life and living. At the very time that biology can and ought to be of greatest service to humankind, we find the biology curriculum stuffed with information that has little potential for application.

Who is most affected? The nonscientist—in other words, almost everybody. We have all heard students comment: "What good is all this stuff going to do me?" We used to get away with the answer that scientific knowledge is justified in its own terms and for its own purposes. That answer is no longer sufficient. Even scientists are facing increasing pressures to demonstrate a return to society from their research.

The thrust today is to bring biology education into harmony with both science and society. The goal must be to help students lead more meaningful, satisfying, and responsible lives, both personally and socially. The curriculum, therefore, must center on human beings as individuals and as social animals. A biology program that encompasses human action may be viewed as a major change because the subject matter of far too many courses today leads nowhere. That is, courses are ends in themselves; there is no expectation of usefulness.

What we must develop is an approach to biology teaching that fully recognizes not only human beings but also their humanity. The most frequent response students give to the question of what they would most like to learn about in a biology course is "myself." Closely ranked on the same question is "to understand people and how to get along with them." (Perhaps we should recall that human beings are the only animals capable of responding to such a question.) But where in conventional biology courses do we find the subject matter to help students discover who they are? Jerome Bruner believes that a major purpose of education is to learn "What makes human beings human; and how can we make them more so?" (Bruner 1966). George Gaylord Simpson, recognizing the duality of human beings and the shortcomings of biological education wrote: "Discussion of man's animality without reference to his humanity or of his humanity without reference to his animality is sterile to the point of inanity" (Simpson 1969). But where in biology is the completeness of human beings taught?

If we examine *Homo sapiens* as an organism, what do we observe? We find the first generation of an organism that is responsible for its own evolution. This animal, different from other animal species, has a brain that is a social organ as well as a physiological one. This brain, with its capacity for learning, has allowed *Homo sapiens* to gain control over the environmental forces that would normally determine the future of a species. Now that we have responsibility for our own evolution, how do we use it wisely to assure our survival at a higher level of attainment than natural forces might be expected to permit? One thing is certain, our answers to that question must determine our perspective on what an education in biology must mean in the 1980s. The primary goals of an education in biology thus become maintenance and expansion of the adaptive capacities of humankind for coping with natural and social forces.

Selecting subject matter for a science of human beings is not easy. While there are gaps in our knowledge, the major problem is that we are drowning in a wealth of information gathered over the past 400 years of modern science. With present-day electronic devices nearly all the information ever known about a topic is recoverable in a matter of minutes or seconds. For example, what is known today about the parts, functions, and interactions of the parts of a cell cannot be recited by a student in a year. (That is one of the reasons knowledge has been called a major threat to education.) In a high school course we have only about 150-160 clock hours to teach a student what is worth knowing in biology. We have long passed the era when that was a reasonable possibility. Yet there is only a small amount of information in today's biology courses that is so basic that—if it were left out—the student's adaptive capacity would be damaged.

The knowledge problem in today's world appears even more complicated when we realize that knowledge itself is undergoing changes in meaning and function. The decade of the 1980s is being described as an information age. The production, storage, and utilization of knowledge rank among our most important national resources. Decisions about whether a country is "developed" are based in large part upon judgment of that nation's capacity to generate and utilize knowledge. There are knowledge-rich and knowledge-poor nations. In the United States, we have expanding numbers of people described as knowledge poor because they lack the knowledge needed to function effectively in our society. Even the knowledge rich look upon some means of artificial intelligence for help in managing what they need to know. We cannot escape the fact that compared to a century ago, a person today needs to know twice as much and in another decade at least three times more to have a minimum education.

There are also knowledge problems in the sciences. The nineteenth-century idea of discipline-bound knowledge, even then only a convenience for data management, is being phased out in the sciences.

Biophysics, biochemistry, environmental psychology, and human ecology are illustrative of emerging integrative disciplines. In all fields of research there is a shading between and among disciplines. The sharpest distinction one can find between disciplines is in the science subjects taught in secondary schools.

The emergence of new concepts about knowledge has major implications for the teaching of biology. What is becoming ever clearer is that too many students are leaving courses information rich, but knowledge poor. These are the students who are the deprived of our society, though they may graduate from a secondary school with a high grade point average. There is a growing number of people in our culture who feel victimized and threatened by those who command organized knowledge and know how to use it. Those who are knowledge poor feel helpless and unable to cope.

Individually and collectively we have more information available than all generations who have lived before us; yet we seem to know less about how to solve contemporary challenges of life and living, as evidenced by our ecological, health, and energy problems. More attention in biology teaching should be focused on ways of making information useable. Biology education must be designed to help students become knowledge rich, in that they know how to obtain information, influence actions, and cope with change.

The new role of biological knowledge in education is its use to extend the adaptive capacity of the human species. That role demands an emphasis upon the personal and social utilization of knowledge and upon the potential application of that knowledge to rational action in life and living.

The utilization of knowledge as central in science teaching is not a new idea. Francis Bacon in 1620 viewed the goal of science to be "for the benefit and use of man, for the relief of man's estate" (Dick 1955). Michael Polyani commented that "true knowledge lies in our ability to use it" (Polyani 1966). Alfred North Whitehead states that "education is the acquisition of the art of the utilization of knowledge." He also refers to science as "an obvious storehouse of ideas for utilization" (Whitehead 1958). John Dewey, at one time a high school biology teacher, pointed out several decades ago that science "in terms of the great human uses to which it may be put . . . offers a field of exploration which cries out from its very emptiness" (Dewey 1960). I do not see how we can continue much longer to deny students the skills required to process and utilize knowledge for their own benefit and for the welfare of the social group.

Turning isolated bits of information into knowledge in a way that is useful for life and living is not inseparable from the discovery of new knowledge; but the processes for acquiring new information are not identical with the processes for knowledge utilization. For the past twenty years, the emphasis in biology teaching has been almost entirely upon developing skills of scientific inquiry, the procedures by which information is discovered and verified. To the research scientist, inquiry skills are of utmost importance as professional tools. These are not, however, the skills of greatest importance when we seek to build personal and social responsibility through the proper use of information already available. Here the need is for decision-making skills—processes for utilizing knowledge to resolve real-life problems and issues. The student who masters skills of information utilization is not only more capable of meeting environmental and social challenges but also more moved to gain additional knowledge as intellectual insights are broadened.

The personal and societal contexts for the use of biological knowledge almost always involve values, often ethics, and sometimes morals. Issues such as human population control, environmental management, and the quality of life are unlikely to be resolved by massing data, however thorough and authentic it is. We have passed the time when science can legitimately be taught in a value-free context. Values provide guidance for the use of information, much in the way that theory provides direction for the discovery of knowledge. Furthermore, the interpretation of biological knowledge in a human and social context has special characteristics: for example, systemic thinking and holistic views in contrast to the linear, analytic reasoning of the so-called "scientific method."

The biology curriculum of the 1980s must also be oriented toward studying and learning the future. Perhaps because biology has always been taught from a historical perspective, students find the subject matter lacks relevance. If biology teaching is focused on life and living, then the subject matter must be chosen for its value in understanding the present and the future. This does not mean the lessons of the past are to be forgotten, but rather taught as they influence the present and guide the future. A future-oriented biology course has meaning for interpreting problems and issues as they exist in life and living not only for today but also for tomorrow. We cannot ignore the fact that life is lived forward—not backward.

The meaning of a future-directed curriculum in biology is sometimes misinterpreted. The most common misconception is that it has something to do with predicting the future. What it really means is that human beings, who are now in control of their own evolution, ought to consider the directions that human experience should take in the years to come. Our vision of the future influences what we do in the present: how we plan, the kind of research we support, the priorities we set, and the decisions we make. In futures teaching, for example, we look forward to some point in time and try to plan ways of building the quality of life to which we aspire. Teilhard de Chardin describes the issue succintly: "The future of the earth is in our hands. How shall we decide?" (de Chardin 1959). Never have human beings been able to influence their future so decisively. Because so many issues about the future are biosocial, biology teachers have a moral responsibility to transform their courses to better suit the possible, probable, and desirable futures of students.

It is sometimes overlooked that science itself is a future-directed enterprise. Through research, regularities in nature are identified which, in turn, allow us to anticipate future events under similar conditions. To be sure, there are sometimes unanticipated results. This is true of all ventures into the future. Scientists respond by seeking more adequate theories or models. Futurists respond by choosing alternative plans.

A biology curriculum for life and living in no way violates the basic facts and concepts of the biological sciences. What it does do, however, is to necessitate the selection of content and learning materials from the full range of sciences that consider human nature and the character of human culture, such as physiology, human or social ecology, botany, zoology, individual and environmental psychology, physical and cultural anthropology, and many more. Each of these fields offers information that can contribute to the development of healthy human beings who can fulfill their obligations to each other, to themselves, and to society. Such knowledge forms the basis of a unified science which links in a holistic way the biological, cultural, social, and psychological aspects of human beings.

There is an evolution in curriculum as in all other things organic.

We are now at a period when the subject matter of disciplines should be viewed as a resource; the actual curriculum must be the context in which the subject matter is taught. The curriculum is a balance of science, culture, and humanity. Culture determines both how we use and modify nature and how we think about it. The new context for biology grows out of human events, the character of the culture, and the concepts of biology that impact most forcefully on human experience.

The laboratory and field work in a biology program for life and living will be different from traditional practices. As with all good laboratory programs, the activities are designed around problems. The difference is that in human biology students not only deal with a problem—they are a part of the problem. Most of the problems are selected from the real world of life and living. Students investigate human behavior, population issues, personal health management, the natural and social environment, genetic profiles, human development, and many other topics. Some of these investigations are individual, but many involve groups of students. After all, people are social animals and biosocial problems demand group actions.

In human-oriented biology, students have opportunities to learn that biological facts are the means of an investigation, not the end. They also learn that most human problems have multiple causes and involve interacting and often uncontrollable variables. Such problems can have logical and rational answers, but are usually not solvable. Like pollution and energy problems, they can be resolved but continue to exist. This is the way it is in the real world of human beings.

What then is the new biology for life and living all about? It is a biology that honors human nature and the character of human culture and that recognizes the cognitive potential of human beings. It is a biology designed to increase the adaptive power of the individual and help assure a favorable evolution of human beings not simply to exist, but to attain an ever richer quality of life.

References

BRUNER, J. 1966. *The growth of mind.* Cambridge: Educational Services Inc.

de CHARDIN, T. 1959. *The phenomenon of man.* New York: Harper Torchbooks.

DEWEY, J. 1960. *The quest for certainty.* New York: Capricorn Books.

DICK, H. (ed.) 1955. *Selected writings of Francis Bacon.* New York: Random House.

POLYANI, M. 1966. *The tacit dimension.* London: Routledge and Kegan Paul Ltd.

SIMPSON, G. G. 1969. *Biology and man.* New York: Harcourt, Brace and World.

WHITEHEAD, A. N. 1958. *The aims of education.* Mentor Book. New York: The American Library and World Literature, Inc.

Commentaries

Hurd's goal is to "help students lead more meaningful, satisfying, and responsible lives." "(A) biology that honors human nature and the character of human culture and that recognizes the cognitive potential of human beings" is what I think biology teaching today should be.

I find myself drowning in a wealth of information as I try to prepare students for achievement tests and college. How much better it might be to help students learn how to obtain information, use knowledge, cope with change, and use problem-solving techniques as a means of learning responsible decision-making.

Jane Abbott

A future-oriented biology course offers the fascinating prospect of continued learning for both teacher and students. The course becomes one of process and not of content. The content becomes the entire range of human endeavors; the process that of responsible decision-making and problem solving. The process requires cooperation between student and teacher, student and student, student and society.

To choose the human organism as the focus for such a course is fortuitous! As Hurd states, most students want to learn about themselves. That desire need not be limited to the biology courses; it should be recognized in the courses students take. Individual students—whether they can state it or not—pursue education for the purpose of enhancing quality of life. Each student seeks "the expansion of the adaptive capacities . . . for coping with natural and social forces."

The human organism is ideal for classroom study. It is always available regardless of the season and requires a minimum of culturing and cleaning by the teacher. It can readily be obtained in both sexes and in a myriad of physiological and morphological types. "Experts" in various fields are eagerly available both from the students, the faculty, and the community.

Biology for life and living is for teachers who are confident in their own abilities and willing to trade the "authoritarian" role for discovery and excitement with students.

Roy K. Baer

The interdisciplinary approach to teaching has gained great respect in recent years. If properly done, this approach yields rewards far beyond expectations. If our lives are not lived as isolated episodes, how can we expect the preparation for our future to be taught in isolated bodies of knowledge?

Hurd notes that the sharpest distinction one can find between disciplines is in the science subjects taught in secondary schools. But isn't that the easy way out for the teacher? Yes, memorize this list for tomorrow's test and get your A! This may make students and parents happy, but it does not prepare a person to deal with problems of the future. Our problems can be faced only with the knowledge of how to find a potential answer, not in the expectation of actually knowing all the answers.

In defense of the teacher, let us remember that to provide such an interdisciplinary approach to biology teaching requires time and planning. Are we willing to pay the price? Hurd answers by quoting de Chardin: "The future of the earth is in our hands. How shall we decide?"

David P. Lapierre

I agree to a certain extent with Hurd that ideas of humanity, science, and technology may be the nucleus on which to build the biology of the future; but traditional topics must still be a part of the curriculum. If quality of life is our goal, we must look to the past for guidance and an understanding of the life around us. The solution may be a new course combining the humanities with science so that answers may be found for the sociological problems being created by our technology and scientific research.

It is time that the social sciences and the biological sciences meet on common ground in the classroom to discuss the problems of the future, using the knowledge of the past. But let's not forsake one for the other; for, if we do, our goals in the future will never be reached.

John G. Lepley

Paul Hurd has expressed in very convincing terms a possible direction for biology education in the 1980s—a direction that recognizes problems and concerns that have been on my mind for a number of years. Hurd recommends that the curriculum should represent a balance of science, culture, and humanity in order to prepare students to use knowledge and develop skills for dealing with problems. No longer can a consideration of social issues be limited to the social studies classroom. Ways must be found to incorporate these issues into the biology curriculum.

My students address these issues in the second year biology course, geared for eleventh and twelfth graders, more often than in the first year tenth grade course, where time is consumed in getting through the "basics." While my personal conviction is that every student should be required to study two years of biology, I realize that convincing administrators (or even other biology teachers) of the truth of this conviction is a "pipe dream." (Perhaps we should not be so timid!) If a two-year sequence is impossible, then a reasonable alternative seems to be a re-evaluation of the existing program, eliminating and de-emphasizing areas that seem less important today than they were in the past.

Possible areas for de-emphasis and even elimination spring to mind immediately. For the past twenty years BSCS has led the way stressing discovery as the best way to teach biology. While this approach is ideal, it is a time-consuming occupation. Is self-discovery no longer the most important consideration? I feel that the needs of the 1980s are greater and more critical than the lessons of the 1960s and 1970s.

The challenge is upon us! New programs must be developed, new textbooks written, new classroom materials produced, teachers must be trained. We must stand up to our convictions.

Betty W. Long

When Andy Lipkis was fifteen years old, he attended a summer camp in California's San Bernardino Mountains. He heard a naturalist explain that the beautiful forest was rapidly being destroyed. "You may be the last generation to enjoy it," the naturalist said. "In twenty-five years, smog will have killed most of the trees, and nothing will be left but barren hills."

"What can we do about it?" Andy asked.

"As long as there's smog, not much," replied the naturalist. "But there *are* some trees that grow in spite of smog."

"Why can't we plant them in place of the ones that are dying?" Andy questioned.

The naturalist shrugged. "You can hardly replant the whole forest."

Andy wasn't convinced, and what followed is the story of one boy's commitment, hard work, and good sense. The result? Andy and other young people are indeed replanting that whole forest.

I was reminded of this story (Nagel 1978) as I read Paul Hurd's paper. Andy's story is just one of many that might be cited to show how development, maintenance, and expansion of the adaptive capacities of an individual can direct motivation and widen capacities to cope with natural and social forces. Individuals so educated are in a position to assume leadership in guiding future human evolution.

Biology for the 1980s and beyond challenges teachers to help students acquire such skills—to help young people gain confidence in their abilities to lead meaningful and responsible lives—both personally and socially. A biology curriculum that recognizes the humanity—and, therefore, the completeness—of human beings can lead students to understand the problems and issues of the future, and perhaps to do something about them. The realization of this goal depends in great measure on how successful we can be in teaching students how to process knowledge and use it appropriately. It also depends on how well they are led to the maturity required for sound judgment and effective action so that they can, with courage and dedication, bring about important changes, as did Andy Lipkis, the "Tree Boy."

Sister Marian Catherine McGrann

Reference

NAGEL, S. 1978. *Tree boy.* San Francisco: Sierra Club Books.

Paul Hurd is most perceptive in recognizing the need for a refocusing of biology teaching. Historically, biology was a "simple" science. It did not require much sophistication in mathematics and could be taught early in a person's education. Even as knowledge in the physical sciences mushroomed, biology remained largely a descriptive science. At the same time that physical science was being divided into chemistry and physics at the secondary level, biology remained in its traditional spot—one year, early in the high school experience. It was not until the 1960s that pressures were felt in biology. Nevertheless, system inertia prevailed, and the science of life remained descriptive in spite of a broad-based attempt to develop an inquiry approach to biology. The new curricula failed to recognize that most students were not going to be scientists, much less biologists. Ignored, for the most part, were those areas of study that nonbiologists need to be scientifically literate in a rapidly changing environment.

Even when the problem became apparent, curricula were added to rather than re-evaluated. Majors and nonmajors alike become "information rich" but they are still illiterate with respect to the human condition. Dr. Hurd's comments poignantly underline this point.

James L. Mariner

Paul Hurd's paper is a thought-provoking assessment of new directions in learning biology. His insistence upon making biosocial information useful to people appeals to me greatly, especially for the nonscience student who seriously and rightly questions learning "all that stuff" in the typical biology class of today. I believe that the basic principles of biology can and should be learned through the interests and concerns of the student regardless of the student's plans. If young people learn *how* to obtain information; *how* human social, political, and economic systems work; and *how* to influence these systems, they will have the coping skills needed for living in their world. Learning activities designed around students' interests and needs can and will be developed and implemented if *human biology* becomes the focus for tomorrow's learner and citizen.

James A. Meyer

Although we are teaching *in* the 1980s, we are teaching *for* the 1990s and beyond. The study of human beings may be looked upon not as a new challenge for biology teaching—but as a renewed necessity. Young children are fascinated with their bodies. As these children grow older, high school teachers and college professors can nurture this inherent interest and expand its scope.

The study of human beings is not limited to the functioning of the organ systems of the body. No matter what the discipline, human examples provide the stimulus needed to gain the students' attention. Once their interest is sparked, students are more receptive to instruction designed to help them develop skills needed to answer questions that will arise in the future. We must emphasize that humans are part of a complex system that includes technology and the various disciplines of biology. If courses are to emphasize a student's future role in society, and I think they should, it must be made clear that diverse parts of the system are linked. While an individual may decide on his or her own future, those decisions affect the future of many others.

What should a student learn about biology to live intelligently and effectively in the world of tomorrow? All people must have the ability to ask questions and should be given information, skills, and tools to answer those questions—perhaps with the help of others. People must have a sense of objectivity—know how to analyze a problem, observe, gather information, and make decisions. Teachers must provide students with these "basics," and then give them opportunities to test their skills before they face problems in the future. Students cannot be taught the solutions to unknown problems, but they *can* be given the equipment with which to deal with unknown situations.

Learning biological facts is not included in the above list of what I consider basic skills. I am concerned that students do, as Hurd suggests, leave our courses information rich, but knowledge poor. I think we must give up some of the immediate returns of our teaching, such as preparing students for the SAT advanced biology tests, so that the *processes* of science can be taught. I am not suggesting that we make everyone into a scientist. Rather, some logical patterns of thinking must be given to students who then, I hope, will use those patterns as guidelines whenever a problem arises.

Gordon E. Uno

2

VALUES AND CHOICES

Edward J. Kormondy

Given the immediate prognosis for change in credulity, leadership, and belief in various social institutions, the most likely candidate to 'do something' is the school.

This is an exciting age of discovery and development in science and technology. But the questions raised, the fear engendered, and the anxiety heightened suggest the public's inadequate preparation to cope with new finds. Advances in biology at the molecular, cellular, organismic, population, and ecosystem levels have significant actual or potential consequences on individuals, society, and the human species. Dealing with such issues in the classroom requires one to plunge into values and beliefs, an area few are prepared to "teach." Human beings are not only capable of choice, but are really incapable of *not* making choices. Choices are based on values. The role of the teacher is to explore possible consequences of choices and to explore issues openly and candidly, without bias or prejudice.

THIS IS AN EXCITING, even exhilarating, age of discovery and development in science and technology. Common parlance about pulsars and quasars, PCBs and polyesters, Krypton and sevrin, recombinant DNA and CAT scans—to sample just a few—strongly indicates the rapidity with which discovery and development become part of the lexicon. Simultaneously, the questions raised, the fear engendered, the anxiety heightened suggest the public's inadequate preparation to cope with these new finds. Underlying much of the concern, but usually not well articulated or comprehended, is the meaning of these advances for self, family, and society both now and in the future. To be told what the immediate effects are of two millirem of krypton radiation may be comforting for the nonce; later may come the more agonizing questions to which there are no answers, questions concerning long-term consequences among others.

Serendipitously, the impact of scientific and technological advances of the past several decades may have an outcome unrelated to the "progress" that inheres in such development. I refer to the stopping short phenomenon, the reaction of wanting to halt long enough to see what's happening and what's likely to happen and to look harder at what really does matter. The compounding element in making such assessments is that too many are poorly equipped to do so since the traditional vehicles for transmitting values have corroded or been corroded to relative impotence. Singular among those forces were the church and the family, and to an only lesser extent, the state and the schools. All these forces have lost their ring of credulity, their force of leadership, their beacon of belief.

Given the immediate prognosis for change in credulity, leadership, and belief in various social institutions, the most likely candidate to "do something" is the school. And, among the least well prepared to do so are the teachers and administrators of the schools. The culprits, however, are the colleges and universities that failed their opportunity and obligation to incorporate ideas of worth and of values and the process of valuing in the preparation of the teachers. Flagellation, however, accomplishes little else than to raise welts and rancor, with perhaps some alteration in behavior. Instead, the response needs to be that of providing guidance and assistance, but mostly encouragement, in dealing with ethical and other values issues in the classroom.

What follows is largely a perspective for reference and a framework for more elaboration and implementation than is appropriate or possible here. First, to prompt the memory, I shall recount some of the developments which have taken place in biology over the past few decades, advances that have significant actual or potential consequences on human individuals and society as well as on the human species. Then I shall discuss some premises about nature, the matters of choice and values, and finally some of the implications all this engenders.

Biological Advances Having Societal Consequences

A roll call of discoveries and developments in the biological sciences in even the last decade or two would not only make for lengthy reading but would doubtless be incomplete. Even more so would be a circumscribed list dealing only with those advances that have direct consequences on the lives of individuals, the welfare of society at large, and the future of the human species. As practitioners of biology teaching, such a complete roster would be a gratuitous offering and thus it seems appropriate to briefly indicate a few of the more prominent of these at the several levels of biological organization.

Cellular/Molecular Level

Perhaps the most exciting yet controversial discovery of the past few years is that of recombining genes through gene splicing and various recombinant DNA techniques. *Genetic surgery* of this sort results, of course, in an altered genetic state, a mosaic of two (or more) different entities or properties, even synthetic organisms. Besides advancing our understanding of how genetic material behaves, manipulation resulting in new combinations of genetic properties enables a directed creation of life forms for specific ends, if you will, a teleological mission. In the summer of 1980, the United States Supreme Court gave legal sanction to this manipulation by upholding the right to patent such "manufactured material." There is thus license for creation of organisms, at present bacterial in organization, which can be designed to address a wide range of contemporary problems such as digesting oil from spills and producing insulin or interferon abundantly and much more economically. By the same token, the license likewise allows for a degree of tinkering with evolutionary and ecological processes far beyond the capacity of any other organism. Misdirected, such intervention could result in consequences beyond the imagination of the best of our science fiction writers.

Yet another kind of genetic manipulation involves conscious direction of a different sort from physical intervention on the genetic material. *Genetic engineering* involves the deliberate intervention in the human reproduction process. The techniques may differ, but the practice has existed since selective plant breeding and animal domestication began. At least one technique, vegetative reproduction, was employed by plants long before they even advanced onto land. Vegetative reproduction is more sophisticatedly referred to now as cloning, the natural phenomenon of identical twinning in humans. While cloning has not yet been achieved in humans, a titillating popular book to the contrary, the likelihood of replicating oneself from one's cells is no longer in the realm of science fiction. Cloning is directed evolution ultimately, perhaps ego-gratifying imminently. But it is no more directed evolution than is *in vitro* fertilization, first accomplished in humans nearly forty years ago but successfully brought through to parturition less than five years ago. Likewise, the selection of sperm and eggs from long-term storage banks provides for selective breeding and hence directed evolution. And finally, less physical but no less technical, a kind of genetic engineering derives from the knowledge acquired about the inheritance of various genetic anomalies. This knowledge has enabled genetic counseling to wield greater influence on the selecting in and out of various components of the gene pool through both prospective and retrospective counseling.

Genetic surgery and engineering give us powerful tools to correct deficiencies in a genome, offset the possibilities of debilitating illness or deformity, increase the access to achieving a greater human potential, and ameliorate extant illnesses and diseases. Applied to crops and livestock, these interventions are employed to enhance productivity and food quality to meet the seemingly ever-increasing demands posed by impending famine and malnutrition. Are these tools too powerful for us at this stage of our cultural and intellectual as well as biological evolution? What benefits with what risks are we prepared to accept? What knowns with what unknowns are we prepared to acknowledge? What choices are we as individuals able to make and what will be made for us by society through its cumbersome decision-making processes—and what decisions will impact on society by the choice made by but one individual or two?

Organismic Level

The development of a considerable array of diagnostic techniques for *prenatal screening* has opened a Pandora's box of ethical and social issues. From such now-primitive approaches as monitoring heartbeat, there now exists a variety of biochemical and hormonal tests, ultrasound, amniocentesis, and even fetoscopy that enable increasing reliability in diagnosing developmental and genetic anomalies. Use of these procedures raises such fundamental questions as the fetus' right to privacy, possibly a constitutional one, and the conundrum of what and when is humanhood and personhood. More likely issues surround the rights of the fetus to life, protection of the defenseless, the entitlement of the mother (and father) to free choice in reproduction and the primacy of individual conscience, and the rights of society with respect to the welfare of its citizenry. With the dramatic rise in teenage pregnancy and largely unwanted parenthood, there is no more poignant nor pertinent issue to be addressed by the middle and high school teacher as well as the rest of society.

The development of literally hundreds of *prosthetic surrogates* from plastic hips and corneal transplants to such substitute regulators as pacemakers and dialysis equipment has not only corrected deficiencies and deformities but in many instances has extended lives well beyond the limits imposed by these imperfections. We have almost reached the point at which virtually any part of us has a surrogate. Besides the questions surrounding the right to alter an inherent life cycle, and the social consequences of ever-advancing mortality's scythe, there is the pervasive issue of personhood. Who am I? Or, am I who I was if I have parts of me that are either artifacts of cultural technology or more parts of someone else?

Intervention and prevention of both birth and death, while not new phenomena from what our diggings into the story of our species convey, have been dramatically enhanced by the development of biochemical and physical and mechanical means undreamed of but few years ago. *Birth control* has reached a high level of sophistication and simultaneously, wide application, through such techniques as suppression of gametocyte production in both sexes by hormonal manipulation, surgical interruption of the passage of gametes, blockage of the entry of sperm into the uterus, and mechanically precluding the implantation of the conceptus. At the other end of life's spectrum, *death control,* not only via the route of prosthetic surrogates but also a wide variety of so-called life support systems, has also advanced markedly in recent years. Prevention of life and prolongation of life are two ends of a spectrum along which ride fundamental issues of rights to personhood and the dignity of the individual, to free choice and individual conscience, protection of the defenseless as well as an entitlement to well being. In addition to these strident ethical concerns, one cannot discharge lightly the evolutionary impact of tinkering with the species' gene pool in the case of birth control and the social consciences of an aging population in the case of death control.

Although the listing of other discoveries and applications could be considerably extended, at least passing mention must be made of the control of human behavior by physical, chemical, and psychological means. *Behavior control* is perhaps elemental in human evolution but the degree to which control can be exercised today surpasses by far the stock and the carrot, or the whip and the candy. Physical intervention by electroshock (or electro-convulsive treatment) to the entire body or by electrostimulation of specified functional areas of the brain using micromanipulatory devices, or psychosurgical removal or destruction of parts of the brain are approaches which are now augmented by the use of psychotropic drugs that serve variously as anti-depressant, anti-psychotic, and anti-anxiety agents. Legal questions concerning the right to privacy of the mind and ethical questions about human dignity and personal autonomy and the right to control one's own body chemistry are fundamental consequences of these kinds of interventions. Even more compelling are the rights to informed consent, a particularly troublesome matter when the individual is suffering from brain damage or derangement, and the fine line between therapy and the social control. What is normal behavior and who decides what it is?

Population/Ecosystem Level

Scientific and technological knowhow has enabled a degree of *agricultural manipulation* well beyond the capacity of mere selective breeding, the process by which modern corn, the large variety of cattle, and the almost innumerable varieties of dogs, among others have been developed. Elaborating on these basically empirical techniques of our forebears, science and technology developed the "green revolution," the conversion or alteration of low-yield crops, a means of providing food for the ever-larger number of ever-hungrier mouths in the world. But the cost of such well-intentioned efforts has been at least twofold: one, the percentage of protein in the high-yield crops as compared to their progenitors is considerably less, and thus protein supplement must be added; second, feeding more people tends to allow for the reproduction of even more people which in turn requires feeding even more people which means reproduction of yet more people—a wonderfully sad example of a positive feedback system. The whole issue of goods and their distribution, the sharing of resources has at its ethical core such issues as interhuman responsibility (is one one's brother's keeper?) and the rights to share equally in Nature's goods, of disenfranchisement because of geographic location, accidents of history, or accidents of race and ethnicity.

Perhaps even more societal consequences devolve from the human capacity for environmental manipulation. The roster of such interventions is so lengthy and well known that mention of but few examples seems sufficient. Our capacity to intervene in processes of the atmosphere (*e.g.*, by adding excessive amounts of carbon dioxide), water (*e.g.*, by excessive nutrients or toxic chemicals), and soil (*e.g.*, by excessive irrigation or removal of top soil) is barely exceeded by our capacity to eliminate or modify habitats beyond their capacity to sustain their natural communities of plants and animals and to eliminate, eradicate, or at least decimate plants and animals by pesticides or poaching. All these activities point to our ability to create tools to destruct. Setting aside that difficult paradox, one nonetheless has to face the consequences of these actions not only on human health and welfare, both now and in the future, but to ask whether defenseless plants and animals also have rights to life. Is, or was, Nature designed for humans who thus are to be regarded as masters of the realm with the right to manipulate it at will for their own gain? Or, is Nature entrusted to humankind who perforce become stewards of the realm? Do we have responsibilities to the yet unborn to provide them something better than we have enjoyed? Are we bound by the tie of justice to those future real persons?

Some Basic Premises

This brief summary of some of the salient scientific and technological developments in the biological sciences and their attendant consequences is, as I noted earlier, neither inclusive nor conclusive. As teachers of biology you can easily lengthen the list and expand on the particular techniques and discoveries. It is, generally, when a teacher begins to consider the consequences of such developments on individuals, families, and societies that there is less assuredness and certainly more controversy. One begins to plunge into values and beliefs, an area few are prepared to "teach." I would argue immediately that one doesn't and shouldn't teach values, but should not avoid the fact that there are values involved. Prior to focusing on that issue, however, it is important to consider some undergirding, but too often not directly articulated, principles about nature.

Time

Perhaps one of the most powerful yet usually not articulated concepts in science is about time. Time can be measured in seconds and milliseconds, in hours or years, but also in eons. In addition, there exists the notion of timelessness, the infinity of time, or eternity. Study of processes over time, be they political, social, or scientific yields two opposing if not paradoxical aspects of those processes: *change* and *persistence.* Among scientific processes involving time, we are aware of changes in climate and physiography (owing to such phenomena as uplift, subsidence, glacial action) as well as of constituent plant and animal communities. Thus towns and cities exist now where once there were glaciers and sea, and mountains now exist where once there could have been cities and towns.

By the same token, there is persistence or a high degree of constancy over time. In spite of drift, the great continents are relatively unchanged in their configuration and outline; particular plant and animal communities persisted over hundreds of thousands of years; climatic trends tend to be of duration measured in thousands of years.

Interactions

Interactions take place through time also. Climatic changes as well as physiographic changes, even if slow and imperceptible, induce change in plant and animal communities. Conversely, given assemblages of plant and animal communities affect local and regional climate and land form.

On balance, Nature is conservative. There is more persistence than change. It is this persistence that renders the scientist's capacity for prediction, for the reliability of scientific description, and the elaboration of constants for this and that phenomenon. This persistence can be viewed as a kind of stasis, a dynamic steady state. Change is secondary but is nonetheless the ultimate emergent over time.

If Nature is conservative, humans generally are not. Although all organisms, microbes, plants, and animals interact with their environment, altering it and in turn being affected by it, humans have capacity beyond the measure of all other organisms collectively to interpose in Nature's conservative posture. We do so by dramatically accelerating the rate of change and more dramatically by directing the nature, degree, and avenues of the change. Clearing of the forests in New England, manipulating the prairies, colonizing and expanding westward, dredging harbors that would ordinarily fill and form deltas, altering stream flow and flooding land by building dams, converting agricultural land to suburban lawns are but a few of a seemingly endless list of alterations. The consequences of these activities are alterations not only in the natural functions of land and water but also of the inhabiting and balanced assemblage of plant and animal communities.

A roster of other kinds of changes can also be developed: genetic alteration and manipulation of the gene pool of both humans and non-humans, control of what to us are pests but which to other organisms may be natural sources of prey or which keep balance among constituent parts.

Who Owns Nature?

This capacity for human intervention into natural processes poses a most fundamental question: who owns Nature? Who has the right to do what to natural processes? Does Nature itself have a right not to be violated?

General western tradition, derived from Christian interpretation of Judaic precepts, holds that Nature was developed for humankind. The long evolutionary process that led to the advent of *Homo sapiens* was focused, pyramid fashion, toward the acme that species represents. In this framework, humans are *apart from* rather than *a part of* Nature. All other animals, all plants, land, water, and resources are for the well being, sustenance, and maintenance of the human species. Exploitation for human gain is not only permissible and allowable but "designed." Nature's sole purpose is to serve humanity.

There is here a concept of limited time and of immediacy rather than of timelessness. It is a hedonistic thought of today, not for tomorrow. "Take therefore no thought for the morrow: for the morrow shall take thought for the things of itself. Sufficient unto the day is the evil thereof" (Matthew 7:34). As we follow such a line of thinking, can we wantonly rape and pillage and alter and manipulate without ultimately destroying the very stuff that enables our sustenance and survival? Can we afford to contravene the long-term adaptive evolutionary processes by destroying the tools that are responsible for such persistence over time? Should not our attitude be one of usufruct, using the fruits of science and technology but not impairing the substance from whence they derive?

Other traditions, among them many Eastern religions and that of many so-called primitive peoples, as well as a dispassionate reading of Earth's physical and biological history, incline one to the view of Nature as a whole ball of wax, each part important to the other parts, and none more important than the other. No one entity is to be sustained at the expense of extinction of another, none manipulated for particular purpose of persistence of any one kind. All parts are ultimately interrelated in the sustaining and persisting of all other parts over time. Each is attuned to allow for the likewise natural process of change to occur. In science, this is a systems point of view, a view that applies to the cell, whole organism, a species, or an ecological unit.

There is a timelessness in this view. In some religions this includes the concepts of reincarnation, of eternal spirits of the departed. This is the eternal recycling of things. There is also here a reverence for life, a recognition of the importance of things for their inherent worth.

In the distinctions just drawn, there is an admitted pushing of the extreme: the issues are not clearly nor necessarily black and white betwixt western and eastern traditions as careful reading of both and their impact will reveal. But for purposes of sharpening ideological polarities, which in fact are upheld by individuals, groups, and societies, the major distinctions of postures on the ownership of Nature may be helpful in addressing the critical issues of values and choices.

Choice and Values

Choice

Human beings are not only capable of choice, but given their unique capabilities and capacities, they are really incapable of not making choices. A hydra responds to touch by reflex behavior; a dog is conditioned to respond to a bell; and a plant's roots grow downward in response to the presence of a hormone. Humans likewise express reflexive and conditioned behavior as well as demonstrating responses to the presence (or absence) of hormones.

But, unlike virtually all other organisms, humans not only can but are forced by their very nature to make choices and, unless there is an impairment of faculties, do make choices. They may be simple behavioral choices such as turning left or right, going forward or backward. They may be complex choices in which social, economic, or other factors are involved. Whatever, choice is an attribute of *Homo sapiens*.

The capacity for choice is often circumscribed by forces beyond the individual's control. Restriction of freedom of choice may be imposed by religious tenets (*e.g.*, thou shalt not . . .), economic constraints (*e.g.*, insufficient funds), political realities (*e.g.*, a one-party system), legal restraints (*e.g.*, incarceration), safety considerations (*e.g.*, 55 mile/hour speed limit) and on and on. In many of these instances, not only has the range of choice been circumscribed, but it has been set by someone else. Whether or not to go to school is dictated by the state through one of its agents. The state made the choice, not the youngster or the parents. But the state is people: thus, the choice is culturally derived. Even with a considerably circumscribed set of freedoms, there remains some considerable latitude to choose.

Values

Why we choose to do what we do depends on what we want to do or have happen. We have a reason for choosing among alternatives, even if it is not clearly articulated or recognized. The reason rests ultimately on what we think is more worthwhile. Values are placed on alternative courses of action, and we choose to do this instead of that because we value the former (or its outcome) more than the latter.

Values are largely derived from culture. They arise from tradition and that, in turn, is conditioned by a number of forces, among the most influential being religion, economics, and sociological elements such as the family, tribe, or nation. Americans tend to value tangible goods (TVs, cars, clothes) more than many non-industrialized peoples for whom the absence of the material makes the non-material more valuable. Although the point may be over-simplified, the dichotomy is nonetheless real.

During the 1960s and 1970s, many Americans and particularly the younger generation withdrew, disillusioned by materialism and overwhelmed by technology and our seeming inability to control its influence and direction over individual lives. They placed higher values on living off the land or communally in harmony and interdependence. Today, the rapidly escalating cost of energy places higher values on fuel-efficient cars, energy-conserving modifications to the home, and others.

A complete listing of the more fundamental of human values would also require more space and explication than this brief essay can accommodate. Nonetheless, one can cite as widely held, but not necessarily universally accepted, values such as: the sanctity of the self and its attendant corollaries of rights to privacy or confidentiality and inviolability, beneficence, utility, justice, and equity. Whether one considers such fundamental moral values as moral imperatives in the view of a Kant (a deontological view) or pragmatically so because of the end results (a teleological view) is less of concern than is the very fact of the pervasiveness of these ethical values.

Some Implications

Survival

Since arriving at the level of humanhood, humans have been at the crossroads of making choices. The values that have been important include survival of the self, family, tribe, nation. Any survival beyond self really implies the survival of the species. Survival may seem to be self(ego)-directed, but in reality it has evolutionary implications, the survival of the species. If this is so, and I contend it is, then a number of questions arise, among them:

- Do we have responsibility as members of a species to the perpetuation of that species? Or, do we accept the story of history that, regardless, some species will persist more or less as is (*e.g.*, the horseshoe crab and Limulus), some will change (*e.g.*, horses), and yet others will become extinct (*e.g.*, dinosaurs). Which fate are we destined for? Such a view requires a perspective well beyond the two or three generations with which most of us have personal acquaintance and extends instead into the hundreds of generations.

- Do all humans have equal rights to survival, or only those who have expressed the will to curtail population size to land and food supply, or those who have shown the capacity to effectively utilize resources, or those of certain intellectual abilities, or those without genetic disease or deformity? Are some people more equal than others?
- Do plants and animals have rights not to be manipulated, maimed, or led to extinction? Do humans have responsibilities to plants and animals because of their defenselessness and their inherent value (an end in itself) or only to ensure human survival (a means to an end)?
- Do soil, water, and air have rights not to be disrupted in their normal processes and rates of such processes because of their inherent worth or so as to ensure human survival?
- Do humans have responsibilities to assure that future generations can also enjoy the fruits and benefits we have enjoyed from Nature? Should we leave this a better place than we found it? Will tomorrow take care of itself?

Law and Choice

Some choices, based on values, may seem beyond one's control as an individual. You may want a clean lake in which to swim and feel powerless to control the flow of human sewage or industrial waste. You may want to halt research using recombinant DNA techniques but feel powerless to act. As individuals we are often powerless, but as individuals banded and bonded together we can affect choice and make choices. Such groupings include governing bodies whose choices, expressed in law or regulation, generally reflect the values of their constituencies. Thus, the National Environmental Policy Act of 1970 and the numerous federal, state, and local laws and regulations affecting air, water, soil, plants, and animals have come into enactment. Similarly, stays on recombinant DNA work were effected both at national and local levels. Society, the collective of individuals, does place values of its choosing on its present and future.

Beyond and because of enactment and enforcement, the courts are called upon to interpret laws and regulations. Like Nature, courts tend to be conservative, preserving the long-term status quo values of society. They not only rule on points of law but conclude on the underlying values expressed in constitutions—equal rights, fair and equitable treatment, integrity, inviolability and rights of the individual. The courts have ruled on the rights of non-human species (snail darters and trees) and of humans (bans on carcinogens) to a healthy and species-survival oriented environment.

You and Choice

In spite of the seeming odds affecting one's stake in the future of humankind, individuals still make choices based on values. To the extent that contemporary biological science and the technology that derives from it pose social consequences, we shall be put in a position of having to make choices and in having to respond to our students (and others) not only with the technical information of how something works or might affect us, but with the very difficult question, "What would you do?" Although that is a legitimate question, it is one that shifts the burden from one person to the other to make a choice whereas it is the individual who must make the choice. Individuals make the beds in which they sleep.

I posit that the role of the teacher is to explore these consequences and issues openly and candidly, without bias and prejudice, and to let the students elaborate the consequences and weigh the effects of those consequences. One cannot "teach" values and shouldn't in the school situation (except perhaps where it is consonant with the particular philosophy of the school as is the case in many parochial schools). But the teacher should assist the student in grappling with the social and ethical implications of whatever belief system they espouse. It's the student's choice in the context of his or her value system. The teacher chooses whether or not to believe in the social consequences of science. But I believe it is your responsibility, and mine, to do so.

Commentaries

I agree that the public in general and students in particular are not prepared to explore adequately the many controversial issues brought about by modern technology and science. Most teachers lack training and, as a result, the courage to explore controversial issues so they can help students learn how to make responsible decisions. Genetic counseling, the right to privacy, interhuman responsibility, or who owns Nature are just a few of the many possible topics requiring decisions.

I was so impressed with Kormondy's paper that I read parts of it to my students. High school students today love a debate but tend to enjoy a discussion "off the top of their heads" rather than studying an issue in some depth. However, in an open, unbiased environment, students do enjoy the challenge of a serious study followed by a realization of the social and ethical consequences of any decision they make. Many teachers, experienced in dispensing knowledge, find it difficult not to try to "teach" values and to keep their prejudice and bias from a discussion.

Teacher-preparation programs should include a solid core of instruction on "teaching" about values and choices. In spite of the current budget-cutting tendency, I feel that a study of values and beliefs (with perhaps a core curriculum developed from Kormondy's paper) should be included in all high schools.

Jane Abbott

Kormondy poses a sweeping challenge to all science teachers: Can we effectively place the technology and knowledge we espouse within a framework for responsible decision-making? The problem is one of pedagogical timelessness but is now compounded by the degree to which the discoveries of the last few decades have shaded many past absolutes from their comfortable black and white to an ever-expanding spectrum of gray.

Choices cannot be avoided even in a community that forbids "tinkering" with the values it lives by and the values it expects youngsters to accept. The issues presented here are inescapable; the technology is now and the decisions must be made now. How they will be made depends upon an informed populace that must understand the issues and respond intelligently. Teachers must help their students understand the discoveries which have been made and must direct them to an understanding of the applications of those discoveries, for themselves and for society. Thus, students must be able to grasp a concept, apply it to their everyday experience, and make reasonable predictions of the consequences of actions. One's values will provide the base line in determining whether such results are "good" or "bad" for the individual and for society.

A prerequisite to such teaching is that the teachers must first clarify the issues for themselves and understand the reasons for their own value judgments. A teacher secure in his or her own values can help others to examine theirs. If we as biology teachers are to "plunge into values and beliefs," then we must understand the discoveries that have been made, the consequences of their applications, the issues that emanate therefrom, and the means by which we can examine our values, beliefs, and the choices that must be made.

Roy K. Baer

Apathy is defined as a lack of interest; a listless condition; indifference. We in education experience apathy each day as we attempt to sell our products to uninterested buyers. Why don't they care? Why aren't they interested in what we are selling? Could it be that we are selling a useless product? Could it also be that many of the leaders of our churches, our government, our schools, and, yes, even our own parents are the products of outdated and useless information? In this exciting age of discovery and development, is the public capable of coping with the many new finds of science?

Our past has shown problems in how we educate our young. We have found that education cannot limit itself to facts, but must nurture the ability of students to make choices based on sound values. The apathy in the classroom may be the result of our attempts to dispense outdated, worn-out, and overused information that no longer supplies an individual with the tools needed to make choices based on values.

Today we have no greater challenge than to deal honestly and clearly with the multitude of biological advances made in the past few years. We must, however, always remember that the individual, when faced with indecision, must finally make the choice. Only when an individual can deal with the consequences of a decision can he or she be considered truly educated.

David P. Lapierre

3

TOWARD AN EFFECTIVE SCIENCE POLICY—HISTORICAL RESUMÉ

Bentley Glass

Every gift of new knowledge that expands human power and enters the fabric of our civilization entails offsetting ills and unsuspected consequences. The greatest need of all, perhaps, in the creation of a farsighted science policy, is to attend with sufficient care to the true costs of new innovations in technology.

Science and technology were not always so closely tied as they are today. Modern science—with its observations, hypotheses, experiments, and theories—is a recent element of human history. In most contemporary societies, science, like the arts, has become recognized as worthy of encouragement and protection; but the history of science policy in this nation is uneven, varying with the political climate of the times and the tone of presidential administrations. Gains have been made through increased funding of the National Science Foundation and the establishment of the White House Office of Science and Technology Policy and the Office of Technology Assessment. But that is not enough; we have not yet created the farsighted science policy that is needed to attend with sufficient care to the true costs of new innovations in technology.

"The greatest challenge which education must face in our time is that of coping with the rapidity of change in science, in technology, in human power, and in the conditions of man's life. The crux of education lies precisely here: that the educated man of yesterday is the maladjusted, uneducated man of today and the culturally illiterate misfit of tomorrow. Education must clearly become a continuing process of renewal" (Glass 1970, pp. 51-52). We are no nearer than we were a decade ago toward incorporating this need into a coherent science policy for the nation.

This paper reviews the history of public policy on science in the United States and outlines steps that biology teachers can take to continue their own growth in the comprehension of modern science and its technological impact on society and to convey this understanding to each new generation of students.

IN 1743 BENJAMIN Franklin gathered some friends together in the city of Philadelphia to form a society "for the promotion of useful knowledge among the British plantations in America." The society thus formed is still with us as The American Philosophical Society. The name of the Society is misleading to many persons, for not many are aware that in the seventeenth century "scientists" were not known as such, but were termed "natural philosophers." Franklin, Washington, and Jefferson, who among others were distinguished early members of the Society, were chiefly interested in *useful* knowledge, growing largely empirically from invention and discovery, and not in basic science that is devoid of any obvious relation to human progress and human power over Nature. Yet Franklin, in particular, cultivated an interest in knowledge not obviously useful, as his interest in electricity and his rather dangerous experiment with the lightning and the kite makes clear. As President of the United States some decades later, Jefferson initiated a concern on the part of the federal government for exploration and the discovery of new and useful knowledge about the vast unknown stretches of the expanding United States, beyond the Mississippi River (Lewis and Clark expedition, 1804-05).

These relations may help to make evident the fact that technology—now so intimately combined with science that each new scientific discovery breeds new technologies that open up new possibilities for the exploration of Nature—was not always so tied to science. The great early technologies—husbandry, farming, mining and smelting of metals, pottery-making and firing, weaving, and working in stone—grew to stature before there was any science, in the modern sense, to assist. Modern science, with its accurate, carefully recorded observations, its dissections and its experiments, its hypotheses and its tests of hypotheses—to say nothing of its broad general theories—is a very recent element of human history, dating back, in the biological sciences, not much before the year 1600. In Franklin's time, agriculture and medicine, the two chief applied branches of biology, were almost unrelated to the science of the day. Agriculture had progressed in the Old World chiefly by the discovery of new food crops, such as the "Irish" potato and the tomato in the highlands of Peru. Medicine had profited similarly, especially by the discovery of the Indians' use of quinine to combat the chills and fever of malaria. For the most part, the medicines in use were folk remedies, such as the colchicine extracted from the yellow crocus of Greece, an empiric that Franklin found helpful in alleviating the pangs of gout and hence brought back to the colonies from Europe. Surgery was still in the hands of barbers with a scanty knowledge of anatomy. Without anesthesia, surgery was excruciatingly painful and often fatal because of faint understanding of how to prevent or control infection. There was no germ theory of disease that was generally accepted. There was no knowledge of the cellular constitution of the tissues of the body. Sexual reproduction and heredity were mysteries.

In the most enlightened countries of the West, science, like the arts, had become recognized by the ruling monarchs as deserving of encouragement and protection. The primary motivation seems to have been national prestige; the mechanism was the foundation and financial support of royal academies. The earliest such academies were the Academia de Ciencias Matematicas of Spain (1575) and the Académia dei Lincei of Italy, to which Galileo belonged. The latter was founded in 1603. The Royal Society of London was established in 1662, and the Academie des Sciences of Paris in 1699. The Prussian Academy of Sciences was established in 1744, and the Academy of Sciences in St. Petersburg in 1724-25. These quasi-governmental academies helped organize scientific expeditions to explore unknown parts of the world, publish scientific papers, and promote discoveries. There was little or no thought that they had any practical, technological value to their countries.

Science, however, was expanding at an exponential rate from its small beginnings at the start of the seventeenth century. Take biology, for example. The number of truly trailblazing or "milestone" discoveries in seventeenth-century biology was only nine or ten. In the eighteenth century there were twenty-two (Glass 1979); in the nineteenth century, 136. The number was doubling about every fifty years. Especially in the latter half of the nineteenth century, many of the discoveries obviously possessed or promised what Franklin had called "useful knowledge." Most, but not all, of the remarkable technological inventions of the nineteenth century were still the product of inventors who were not a part of organized science. Mechanical inventions and power machines might originate from artisans and mechanics, but practical chemistry and geology were prevailingly the product of professional scientists. Some of these discoveries were clearly of importance in warfare, and this dawning recognition led President Lincoln and the Congress of the United States to establish, in 1863, a National Academy of Sciences of the United States. Its primary function was to provide free advice to the government of the United States upon all scientific matters. Thus, belatedly and under the pressure of wartime needs, the United States finally followed the example of the European monarchs in the promotion of science.

That move toward a national science policy seemed to be enough for a long time, even though the natural sciences underwent a tremendous increase during the twentieth century. For example, the new science of genetics, dating from the rediscovery of the laws of Mendel in 1900, has undergone an exponential increase amounting to a doubling of the most highly significant discoveries in each successive period of 22½ years (Glass 1979). Many of these discoveries had practical applications of vast consequence for animal and plant breeding, health, and medicine. Thus, hybrid corn, a product of the insight and ingenuity of George Harrison Shull and Donald F. Jones, alone increased the yield to corn farmers by 20 percent during the years of World War II—a gain of 1.8 billion bushels worth more than $2 billion at current prices, and enough to pay for the Manhattan Project (Mangelsdorf 1951; Glass 1980). In the area of health, it helps to note that the National Institutes of Health, a major present investment of our federal government in the biomedical sciences, began as far back as 1887 with the establishment of a Laboratory of Hygiene that, in 1891, was moved to Washington, D.C., and renamed the Hygienic Laboratory. By 1930 the National Institute of Health took shape, to be joined by the National Cancer Institute in 1937, and together moved to Bethesda, Maryland, in 1938. Thus, research in the biomedical sciences took its place in government-sponsored research alongside such older agencies as the Geological Survey, National Bureau of Standards, and the Fish and Wildlife Survey. In agriculture, too, a considerable amount of practical and even pure research went on in the U.S. Department of Agriculture Experiment Stations which were mostly located at the land-grant colleges and universities.

During the 1930s, the growing cloud of totalitarian political beliefs introduced new relationships between science and government. On the one hand, Adolf Hitler reasoned: "Science is a social phenomenon and, like every other social phenomenon, is limited by the benefit or injury it confers on the community. The idea . . . of free and unfettered science is absurd" (reported by Hermann Rauschning). This theme was elaborated by the British physicist J. D. Bernal in *The Social Function of Science* in 1939. In the Soviet Union, the rise of the agriculturist T. D. Lysenko to supreme power over Russian biology and agriculture led to the death or exile of a number of notable geneticists and the removal from office or research positions of many more. It quickly became clear that the freedom and independence of the science enterprise could become threatened, deformed, or even killed by too much governmental control. Some scientists, to take the opposite extreme, were pushing the opinion that scientists should not meddle in morals or politics. In the words of the noted British physiologist A. V. Hill, expressed in 1933:

> Science should remain aloof and detached, not from any sense of superiority, not from any indifference to the common welfare, but as a condition of complete intellectual honesty. . . . If science loses its intellectual honesty and its political independence, if—under communism or fascism—it becomes tied to emotion, to propaganda, to advertisement, to particular social or economic theories, it will cease altogether to have its general appeal, and its political immunity will be lost. If science is to continue to make progress, it must insist on keeping its traditional position of independence, it must refuse to meddle with, or be dominated by, divinity, morals, politics, or rhetoric. (Glass 1962)

World War II brought about a critical change in attitudes. A. V. Hill himself was soon occupying a government post, and many other British scientists became engaged in the wartime struggle for survival in the face of Nazi attacks. In the United States, it was the appeal of scientists to Franklin D. Roosevelt that launched the secret Manhattan Project and produced the atom bomb. Openly, an Office of Scientific Research and Development was established, with Vannevar Bush as its able head. Novel weapons such as the proximity fuse, and novel systems of defense such as radar were developed. Biologists were involved in these and other wartime scientific research agencies, although, of course, in far fewer numbers than physicists or chemists. The Office of Naval Research supported considerable biological research. The era of organized research and development had begun.

All the wartime developments took place without any specific legislation or presidential declaration that could be taken to signify a definite science policy on the part of the American government. After the war, these matters came to a head. First was the struggle to establish an Atomic Energy Commission that would place with civilians, rather than the military, the charge to develop atomic energy for both wartime and peacetime purposes. In 1948 a National Heart Institute was established and combined with the previously existing National Institute of Health and National Cancer Institute to make up the National Institutes of Health. And in 1950 the policy of supporting pure as well as applied science by means of federal grants and contracts took shape in the form of the National Science Foundation. The system of approving for support proposals that had been screened not only by staff members but also by peer review panels of non-government scientists commenced, a system that was widely adopted (although not universally) by the other government agencies.

The scientific world was shocked when, within months after the United States had unveiled its hydrogen bomb, the U.S.S.R. detonated its first superbomb. An even greater political and scientific panic swept America, however, when in 1957 the Russians launched Sputnik I, the first orbiting satellite. Millions of Americans asked how it was possible that Russia had outstripped the United States in space science. President Dwight D. Eisenhower had not been greatly concerned about science or science policy, but Sputnik and the American reaction to it changed matters abruptly. Funds for the National Science Foundation were increased by nearly three times in the ensuing budget. The Department of Defense (DOD) was reorganized, with greatly increased emphasis on the support of technological advances in military and paramilitary matters, and a DOD Director of Research and Engineering was appointed. In 1958 the National Aeronautics and Space Administration was established and the National Defense Education Act was passed by Congress. Many other measures were undertaken to ensure that adequate education and training in science and engineering would provide the corps of future personnel to compete with the U.S.S.R. In the fear that our deficiencies began at the youngest ages of science learning, novel science education programs were begun, first in physics and mathematics, and a bit later in chemistry, earth and space science, and biology. The Biological Sciences Curriculum Study (BSCS) was the direct and most widely used and successful of these curricular developments, bringing scientific methods of inquiry and investigation into the high school and even lower grades of schooling and updating the woefully antiquated substance of what was being taught.

Perhaps the most important of the new measures was the elevation of the President's Science Advisory Committee (PSAC) to full status in the White House organization and the designation of its head as a special assistant to the President. As Lee DuBridge, himself later to hold that position, said at the time: "This committee has quietly and effectively rendered enormous service in bringing to the President, the Cabinet, and the National Security Council the best scientific advice in the country on problems of national defense, international relations, public health, education, and the encouragement of research." James R. Killian Jr., was followed by George Kistiakowsky, and he in turn, upon the election of John F. Kennedy, by Jerome Wiesner. Wiesner, like Kistiakowsky, had been a member of the Pugwash Conferences on Science and World Affairs that took place in Vermont in September of 1961 just after the Russians had resumed atmospheric weapons testing and the prospect for peace looked bleak. Viewed with grave suspicion by certain members of Congress during the Cold War days of the McCarthy witchhunts, the Pugwash Conferences offered an unparalleled opportunity for scientific leaders of the Western countries to confer with leading Russian scientists and others from the Soviet sphere of influence regarding the most hopeful ways to preserve peace and avoid the proliferation of nuclear weapons. Many of the suggestions for joint international undertakings in scientific and economic realms that President Kennedy proposed to the United Nations in the spring of 1962 came directly to his attention through Wiesner, who paid attention to the recommendations of the Pugwash Conference.

No president of the United States exhibited a greater sense of the importance of a sound science policy than did John F. Kennedy. He did not live to see brought about much of what he hoped for in this area of government; but his great goal for achievement in space science and technology, that of putting a man on the moon by the end of the decade, became a reality quite on schedule. The massive growth and development of the NASA programs required to bring this about were initiated in the 1960s.

The problems of the Vietnam War embroiled Presidents Johnson and Nixon more and more fully, and matters of science policy and funds for science took on a minor role (Bronk 1974; Kistiakowsky

1974). The influence of the President's Science Advisor declined. The once powerful PSAC was curbed, and its advice was no longer sought. In part, this may have been because PSAC represented, or was thought to represent, too narrowly the scientific and technological interests of the strong Harvard-MIT group of physical scientists and engineers. Biologists had never been well represented on that influential Committee, and no President's Science Advisor had been a biologist. Rather, they had been physicists, chemists, or engineers. In part, too, the decline of power of the office lay in the opposition of Edward E. David, the sixth of the President's Science Advisors, to a favorite Nixon scheme, the development of a supersonic transport plane. In any case, the advisors close to Nixon undermined the power and curbed the functions of the Office of Science and Technology and of the President's Science Advisor. For example, John D. Ehrlichman is quoted by Kistiakowsky (1974) as saying, "(We) needed no policy advice, only facts." And Henry Kissinger wanted to have all technological security matters filtered through him. In 1973, PSAC, the Office of Science and Technology, and the position of the President's Science Advisor were swept away, not to be revived until the presidency of Gerald Ford. Then a reorganization took place, and the White House Office of Science and Technology *Policy* (OSTP) was established with a President's Science Advisor to head it.

The influence of the President's Science Advisor did not become strong again until the administration of President Carter. Meanwhile, environmental pollution, the exhaustion of America's energy supplies, and our growing dependence upon foreign oil caused technological crises to become critical. Today, the need for a wise, farsighted national science policy is even more apparent than it was in the Sputnik era.

If we compare the 1979 governmental commitment to scientific research and development with that of twenty years ago, we can readily see how much has happened. The annual budget of the National Science Foundation has increased to nearly one billion dollars, from $136 million in 1959. Even making allowance for inflation, that is a tremendous increase, and one matched by the sums for research and development allocated by the Department of Defense, the Department of Energy (replacing the Atomic Energy Commission), and the National Institutes of Health. In addition, there are new agencies operating now in science and technology that did not exist two decades ago. Among these are the Environmental Protection Agency, the Office of Technology Assessment, and the Arms Control and Disarmament Agency. The total budget for research and development is now about $30 billion a year, of which about $4½ billion support basic research. Never in the history of civilization has there been such a far-reaching undertaking on the part of government to support the technological basis of progress in material affairs. This has come about through a realization that in the twentieth century, scientific discovery, instead of being a mere cultural delight, is the essential soil and root of all technologies. America continues to lead in industrial and agricultural production, in business, and in commerce not so much because of our more abundant natural resources, but rather because we lead other nations in scientific discovery and application. Space exploration, super-computers of ever-diminishing size, genetic engineering to produce rare and highly useful agents of biological growth and health, and possibly new sources of inexhaustible energy—these are exciting new frontiers.

Nevertheless, as I have pointed out on a number of occasions (Glass 1970, 1971, 1976, 1977), every new advance in technology creates long-term effects and side-effects that subtract substantially from whatever the immediate benefits may be. The Greeks had already grasped that truth when they tied together the myths of Prometheus, the giver of good gifts to mankind, and Pandora, who released unwittingly the plagues and ills that beset humanity's physical and moral existence. Every gift of new knowledge that expands human power and enters the fabric of our civilization entails offsetting ills and unsuspected consequences. The greatest need of all, perhaps, in the creation of a farsighted science policy, is to attend with sufficient care to the true costs of new innovations in technology. Our establishment of the Office of Technological Assessment by Congress, under the spur of Emilio Daddario, was a hopeful first sign of an awakening to the public need in this respect. But sad to say, the Office of Technology Assessment gives no great promise of achieving its goals. It is understaffed and underbudgeted, and can scarcely begin to analyze and deal with the thousands of problems that require examination. A country such as ours, built on successful technology, needs enormous laboratories rivaling the National Institutes of Health to deal with each new technological introduction with reasonable promptitude. Otherwise we will continually be subjected to new perils of the same nature as the thalidomide tragedy in the drug field, the rampaging effects of chemical pesticides and pollutants, and the insidious effects of ill-advised dams, unmonitored additives to foods, and the combined consequences of radiation and chemical wastes released into the air and water. A real part of the problem is that responsibility is fragmented between agencies that fight for priorities and prestige. There is little cooperation between such enforcement and protection agencies as the Office of Technology Assessment, the Food and Drug Administration, the Department of Health and Human Services, the Nuclear Regulatory Commission, and others. The picture is one of confusion engendered by the lack of a clear science policy at the level of the President and the Congress.

One final aspect needs comment. As I wrote in 1970:

> The greatest challenge which education must face in our time is that of coping with the rapidity of change in science, in technology, in human power, in the conditions of man's life. The crux of education lies precisely here: that the educated man of yesterday is the maladjusted, uneducated man of today and the culturally illiterate misfit of tomorrow. Education must clearly become a continuing process of renewal. (Glass 1970, pp. 51-2)

We are no nearer than we were a decade ago toward incorporating this need into a coherent science policy for the nation. Indeed, if the Department of Education is dismantled as President Reagan proposes, we may be a lot farther from it than we were.

For some years during the 1970s I organized and offered at my university a course entitled "Science and the Future of Man." It was planned to meet, on a local level, the needs of which I speak. But no one was found to carry on such a course when I retired. Jacob Bronowski recognized this need when he organized his superb film program *The Ascent of Man.* The educational problem to be faced, however, is that of making such individual efforts nationwide on a continuing basis. Our one-shot curriculum reforms of the post-Sputnik era have mostly died away. Only the BSCS continues, and it lacks the firm support it once received from the National Science Foundation. Sporadic efforts in curriculum and course content reform are not enough. Science policy in the educational field is today almost nonexistent. We now recognize our needs. Can we generate far-reaching science policies and the coordinated mechanisms necessary to put them into effect? The impending disaster that will result from inaction leaves us very little time.

What can biology teachers do to meet these urgent needs? Collectively, of course, we can continue to bring political pressure to bear on Washington for the formulation of a truly farsighted science policy. Especially, the National Science Foundation needs to revise its "off-again—on-again" policy of support for curriculum development for the secondary schools. The crisis of Sputnik is a *continuing* thing. Already, since that date in 1957, our fund of scientific knowledge has virtually *quadrupled,* and the state of the biological sciences so well reflected in the first BSCS textbooks and laboratory programs is now as antiquated as the biology of 1900 was antiquated in 1957. The job is a *continuing* one and must be redone in every decade if our scientific-technological society is to be based firmly on a people with adequate knowledge to vote for a sound science policy. A democracy simply cannot continue to function on the basis of a tiny scientific elite, like a mighty pyramid balanced perilously on its apex.

The individual task of the teacher is twofold: to continue one's own growth in the comprehension of modern science and its technological impact on society; and to convey this understanding to each new generation of students. As to the first, self-education through reading and study is probably insufficient. We need a widespread network of courses dealing with "Science and the Future of Mankind" in our college and community programs of continuing education, and every teacher needs periodically to take advantage of such programs of continuing education. But the most imperative task is to translate our current biological knowledge into social terms in the classroom and laboratory, so that young students may awaken to the imperative need to know what biology can tell us about the growth of science and technology, the foundations of healthy nutrition, the lengthening life span, and the problems of old age facing us as individuals and as a society; the biology of sex and family relations; the biological control of population increase and the possibility of ensuring new generations freer of genetic defect; the biological basis of better physical and mental health and the new promises of medicine; the fateful biology of the consequences of nuclear war; the hopeful biology of the control of animal and plant pests without the ruin of the environment; the biological consequences of the chemical pollution of the environment; the ultimate dependence of humanity upon renewable sources of energy, among which biological sources must play a great part; and of course the humanistic ethics that must guide us as we face hard choices in these and many other matters. In a very real sense the biology teacher has become an important trustee of humankind. The biology we teach must face the future, as it examines the present in the light of the past. Human evolution is coming under our own control—directly as we face the challenge of genetic engineering, indirectly as we face the pressures of the environment we so heedlessly modify.

Biology teachers so often think only of the bright students they may guide into future biological pursuits. That is too shortsighted. We can do very little to reeducate our presidents and congressmen in order to develop a sound governmental science policy. But every future president and congressman is a onetime student in our schools, and nearly all of them will have taken high school biology. Here, as nowhere else, we can find our opportunity and our calling.

References

BERNAL, J. D. 1939. *The social function of science.* London: Routledge & Kegan Paul; Cambridge, Mass. and London: MIT Press, paperback ed., 1967.

BRONK, D. W. 1974. Science advice in the White House. *Science* 186:116-121.

GLASS, B. 1962. Scientists in politics. *Bulletin of the Atomic Scientists* 18(5):2-7.

———. 1970. *The timely and the timeless: The interrelationships of science, education, and society.* New York and London: Basic Books.

———. 1971. Prometheus and Pandora—1971. *Bulletin of the New York Academy of Medicine* Second Ser. 47:1045-1058.

———. 1976. Ethical considerations on the applications of biology. *Anticipation* 22:44-46.

———. 1977. The scientist: Trustee for humanity. *BioScience* 27:277-278.

———. 1979. Milestones and rates of growth in the development of biology. *Quarterly Review of Biology* 54:31-53.

———. 1980. The strange encounter of Luther Burbank and George Harrison Shull. *Proceedings of the American Philosophical Society* 124:133-153.

KISTIAKOWSKY, G. B. 1974. Presidential science advising. *Science* 184:38-42.

MANGELSDORF, P. C. 1951. Hybrid corn: Its genetic basis and its significance in human affairs. *In* Dunn, L. C. (ed.) *Genetics in the twentieth century.* New York: Macmillan.

Commentaries

Glass has hit upon the area that teachers have historically been reluctant to enter. Teachers, being content to leave politics to the politicians, have paid a price for their lack of involvement. The anti-intellectual tenor of the times must surely have alerted everyone involved in education to the necessity of becoming actively involved in the decision-making process at the political level. Of necessity, the era has ended when teachers can close the classroom door and teach without regard for the myriad social and political influences that are struggling to control what and how they teach. Glass' analysis points out the reality that decisions made at the highest levels of government directly affect teachers in the classroom. We must have some influence upon that process if we are to protect and promote what we know is sound science teaching.

Roy K. Baer

Science policy today is captive in the magnetic field of finance. Change in policy can be accomplished only when a dollar value can be attached. Environmental education is supported until the environment (wooded study area) is found to be more "valuable" as a parking lot for the sports arena to fatten the athletic coffers.

If, as Glass so aptly states, education must become a continuing process of renewal, how does one bring about effective policy change—when policy is governed totally by financial need and promise? Our only hope lies in the selling done by the educators of today—the selling of basic needs, not only financial, but of life itself. Science educators must sell scientific discovery as the essential soil and root of all technologies.

David P. Lapierre

Dr. Glass has effectively traced the history of conveying "useful knowledge" of inventions and discoveries in Benjamin Franklin's time to scientific discoveries and new technologies of today. This account vividly shows the social, political, and economic implications of these discoveries, inventions, and subsequent technologies. His observation that we "lead other nations in industrial and agricultural production in business and in commerce . . . because we lead other nations in scientific discovery and application," may not continue if we do not devote more of our gross national product to research and development.

America is in critical need of a coherent, long-range science policy. Cutting back on expenditures for science education can seriously impair our ability to inform all Americans of the benefits and risks of current and emerging technologies which affect all of our lives. I agree with Dr. Glass that we must act *now* on this urgent matter of a coherent, long-range science policy.

James A. Meyer

4

SCIENCE AND PUBLIC POLICY

Gilbert S. Omenn

If we look back twenty years, we see that remarkable achievements have been made. I see no reason to be less optimistic about the future.

The Office of Science and Technology Policy headed by President Carter's Science and Technology Advisor, Dr. Frank Press, was involved in four major initiatives in which biological science had a large interest. These four were basic research, innovation, international aspects of science, and regulation.

During those years, there was an increased emphasis on basic research in human and non-human biology. The Carter administration perceived an overwhelming need for new knowledge and for the application of that knowledge on a sound basis. For example, the country was thought to be far short of its future capabilities in medical technology.

Regulation was an important political issue. The federal government felt that it had been successful in protecting health in the workplace and in the environment, but much litigation was pending. The adversarial relationship is not a comfortable one; but scientists must be willing to provide interpretations of data, to explain what is known, and what needs to be learned in order to develop environmental control measures that best serve the people.

It is important that our country, like other leading countries, recognize that education is a vital public investment. Complicated topics in fast-moving scientific fields are a challenge to educators, citizens, and students.

IN MARCH OF 1979, President Carter issued a Message on Science and Technology to the U.S. Congress. The first two paragraphs are as follows:

> This year the world marks centennials associated with two great men of science and technology. Albert Einstein, who was born March 14, 1879, enlarged our knowledge and understanding of the universe and changed the way we look at space and time. He was an early

> pathfinder in the remarkable era of scientific exploration in which we live. In that same year, Thomas Alva Edison invented the electric light. He was a great technologist who devoted his life to the creation of new products to meet human needs.
>
> The world was changed by the work of these two men. The vast changes that have grown out of their work are part of the reason that the forces these two men represent, science and technology, have become increasingly important in our lives over the last century. We look to the fruits of science and technology to improve our health by curing illness and preventing disease and disability. We expect science and technology to find new sources of energy, to feed the world's growing population, to provide new tools for our national security, and to prevent unwise applications of science and technology. The health of our economy has been especially tied to science and technology; they have been key factors in generating growth, jobs, and productivity through innovation. Indeed, most of the great undertakings we face today as a Nation have a scientific or technological component.

To show how thoughts like these can influence public policy, I will focus on four major initiatives of the Office of Science and Technology Policy in which the biological sciences, during the Carter administration, had a large interest. These four were basic research, innovation, international aspects of science, and regulation.

First consider basic research. It was Lyndon Johnson who said that we must unlock all those secrets that are being kept in the laboratory. One of the reasons that biomedical research has been so handsomely supported by the Congress is that applications have been expected. Some remarkable gains in medicine have indeed been achieved, and promises have been very expansive. The apogee of this approach, of course, was the "war on cancer," from which a dramatic solution was expected. That rhetoric, which lumped all cancers together and seemed to promise a single therapeutic or preventive approach, was not well-founded biologically. It did generate huge increases of funding for cancer research, much of which was very well spent and generated new knowledge, important to cancer and to other fields as well. However, the emphasis on cancer research generated a backlash and a certain amount of jealousy from people working on other medical and biological problems.

In the late 1970s, there was a striking shift in emphasis in the National Institutes of Health and in the National Science Foundation toward individual, investigator-initiated projects—most of which could be described as basic research in human and non-human biology. In the fields of cancer, cardiovascular disease, arthritis, digestive diseases, neurological diseases, and mental illness, it is clear that we cannot stand pat with our present knowledge and simply refine our delivery systems. Rather, there is an overwhelming need—for both humanitarian and scientific reasons—for new knowledge and for assessments of the applicability of our knowledge to clinical challenges for which we have so little now to offer.

Lewis Thomas' books, *The Lives of the Cell* and *The Medusa and the Snail,* offer positive reminders about the history of science in this century. Too often in our impatience, I believe, we forget that only twenty-five years ago there was no coronary bypass surgery or even open-heart surgery; thirty years ago, there was a raging argument whether proteins or nucleic acids were the genetic material. The list of remarkable and very recent changes in our scientific knowledge and in our medical, agricultural, and environmental capabilities is very long. I am optimistic about what our knowledge and technology will be in the year 2000. That year may seem far away, but it is less than twenty years from now. If we look back twenty years, we see that remarkable achievements have been made. I see no reason to be less optimistic about the future.

The second topic—innovation—provides a striking contrast with policy for basic research, because, in innovation, the role of research and development is just one piece of a complicated matrix. This topic involves a host of issues far beyond direct investment in research and development. Some of the larger issues are patents, procurement, tax policy, venture capital markets, international competitive factors and balance of trade, jobs, and productivity. The matrix is larger still because there are many other factors outside the influence of government altogether. These include the response of the industrial sector to public and congressional demand for health, safety, and environmental regulation. In many of our most productive industries, the practice of dividing large companies into small profit centers leads to severe cost-cutting. That usually means pushing all the research and development out of the line divisions of a company into an isolated research branch.

There are aspects of innovation, of course, in the biological and biomedical fields. One involves recombinant DNA techniques, one tool among many in modern genetics for somatic cell alteration. These techniques will have broad application in agricultural, chemical, and pharmaceutical fields. Innovation is needed also in the broad area of medical technologies. There is a good deal of rhetoric (much of it very appropriate) about the dominance of hospital-based, "halfway" technologies to deal with medical problems arising mostly in the last year of patients' lives. These interventions are extremely expensive in dollars and in the emotional burden on the patient and the family. And they are too often unsuccessful. While we want to ensure that these technologies—including coronary bypass surgery, CAT scanners, renal dialysis, and many others—are available for those patients who can benefit, we must also be sure that they are not wasted on patients for whom they may be harmful, not beneficial, or merely too little too late. The other side of the medical technology story that is too seldom discussed is that we are far short of our future capabilities in dealing with most medical problems. We must stimulate innovation directed at interrupting and reversing disease processes through understanding of the pathogenesis of diseases and the promotion of good health.

The third area for policy formulation is international. In this domain, a number of countries including the United States, have recognized that science and technology are closely related to the larger political and economic issues of our times. The Carter administration worked with Japan, for example, to arrange joint funding of some of the largest projects in such areas as fusion research and coal liquefaction. Such projects are so expensive that no country can easily appropriate all the monies required. Cooperation is indicated because the results of the work can have global benefits. The new regime in China provides another opportunity. The political and philosophical themes of the past have been replaced by policies of economic pragmatism, academic competition, academic excellence, and reliance on science and technology in virtually every sector. The Chinese look to us for leadership and cooperation.

In 1979, the World Health Organization officially celebrated the eradication of smallpox from this planet. This disease has been dreaded throughout human history. It is a disease for which an effective intervention, vaccination, was developed in 1798. But it took a long, long time until vaccination was widely used. The delay came about, at least in part, because there was no knowledge whatsoever about viruses, let alone the virus of smallpox. Sometimes we can make practical gains in the absence of detailed understanding. But, in the absence of understanding, there is greater resistance to the application, and the application can be misunderstood. That was true also in the case of polio vaccine. Polio vaccinations were attempted in the 1930s and failed, because no one knew that there are three different kinds of polio virus. The process of vaccination against polio was discredited until more knowledge was available and bolder individuals took up the challenge with trivalent vaccines. The United States is proud of its contributions to the World Health Organization's fight against smallpox. Three hundred U.S. Public Health Service officers served in the worldwide campaign over the years, and the Agency for International Development and the Public Health Service contributed many resources.

My point is that we may see a growing involvement of scientists from our country in the resolution of problems of the developing countries. Study of such topics as acid soils, schistosomiasis, and the feasibility of various population and family planning measures must be done in the places where programs are to be carried out.

The fourth area of special interest is regulation. There is no hotter political item in the country than "regulation." It has come to serve as a code word for everything that the government does wrong. Yet environmental and safety regulations have had important successes, such as cleaning up lakes and rivers, reducing air pollution, and protecting health in the workplace. There is a great deal of confusion about environmental regulation, and part of the reason lies in the differences between lawyers and scientists. Most of the people involved in litigation and regulation are lawyers and other non-scientists. They live on a diet of confrontation and advocacy; the adversarial relationship is comfortable for them. It is generally not comfortable for scientists, although scientists contribute to the controversy when they make differing interpretations, often for very good reasons, of similar data. That has happened in the cases of fluorocarbons, lead, saccharin, nitrites, ozone, and many others. Experts do disagree about the evidence, which in some cases is meager. Our approach should first be to try to increase the amount and improve the quality of scientific research. Second, leaders of regulatory agencies should have a standard procedure whereby appropriate expert scientific advice can be brought to bear in evaluating the options presented.

Perhaps the person who has described this need best is Judge David Bazelon, a distinguished jurist in Washington, chief judge for thirty years of the U.S. District Court of Appeals. He has written and spoken about what he calls the "perils of wizardry"—that is, the dangers of experts extrapolating far beyond the area in which they have expertise. He has suggested that courts that see expert witnesses use the following approach: instead of allowing or asking an expert to advise the court on what to do (which is a judgment in which many factors must be taken into account), courts should have experts testify what is known, on what basis they conclude it is known, what is not known that is important to the decision, and what over the next few years might be learned. The result would be that the record of the court proceeding would have many elements of an equation, not just an answer. Because issues tend to recur, it is very important to build a record that allows re-examination and readjustment of each element of an equation rather than just asking the global question all over again.

Challenges for Science Education

All of these matters have important implications for education in general and science education in particular. The Department of Education was established to dramatize the fact that education is one of the most important aspects of our society. Education should not be taken over by the government, but I think it is time that our country join all the other leading nations of the world in recognizing education as a matter of public interest so important that it should be represented in the highest council of our government. About 10 percent of the science education program of the National Science Foundation has been transferred to the Department of Education. That part concerns teacher training, dissemination of materials, and other activities related to the broader aspects of education. The scientific and curricular programs remain at the NSF.

Biology teachers face an awesome challenge. The news magazines have declared that we now live in a "biological revolution" and that we are enveloped by the "new genetics." Whatever the catch phrase, there is no doubt that genetics, the neurosciences, and related topics have captured the public imagination. Because these are complicated and fast-moving fields of inquiry, it is a challenge to teachers, students, and citizens to learn, to understand, and to use this information well. There is no end to controversy in a rapidly moving field. There are controversies about scientific claims and discoveries. There are controversies about methodology and applications. For example, we know that there are potential hazards in the uses of radiation in medicine and in nuclear power. How great the hazards are and what can be done to minimize them are the subjects of tremendous debate right now. In fact, the Report by the President's Commission on the Accident at Three Mile Island dealt extensively with human factors, human behavioral variables, approaches to assessing and monitoring radiation, biological effects, health effects, and prophylaxis against exposures.

I recently read an article about the teaching of controversies in the classroom (BSCS 1978). As I understood it, the authors were advocating considerable discussion about the political and social aspects of these controversies in the time allotted to teach science. Rosenthal (1979), after reporting what appears to be a good curriculum for a science and society course for high school students, pointed out that very few students take the course. I would raise a caution. I am not enthusiastic about most current teaching on "science and society." For example, I think that the teaching of genetics and of somatic cell techniques, including recombinant DNA, offers plenty of opportunity for introducing limited elements of the public debates as part of a sensitizing discussion. But the science is complicated, and students deserve an opportunity to learn the science in depth. We should not deflect too much of their time and the biology teacher's effort into what are essentially social and political dimensions. Given the rapidly increasing roles of technology in our society, I believe that time might be found in political science and social science classes to address in depth the political aspects of such technologies. Those teachers need the students and their fellow teachers to bring good scientific knowledge to these discussions.

Changes in science and in technology influence and challenge the beliefs of our society. I think scientific issues will become increasingly important in public policy. Citizens in general will require firm grounding in science. Teachers of biology have a grand opportunity to stimulate young people into careers in science and in the appropriate uses of science and technology.

[*Editors' note:* In 1981, the Reagan administration gained approval for initiatives that eliminated federal support for science education and placed the Department of Education in jeopardy.]

References

BIOLOGICAL SCIENCES CURRICULUM STUDY. 1978. Issues and controversy in teaching science. *BSCS Journal* 1(1):33-37.

ROSENTHAL, D. B. 1979. A "science and society" course for high school students. *American Biology Teacher* 41(6):336.

Commentaries

Omenn wisely advises science teachers to stick to the science of issues and avoid political and social controversies. Omenn quotes Bazelon on the "perils of wizardry"—that is, the dangers of experts extrapolating far beyond the area in which they have expertise. I would remind all biology instructors of the confrontation in which Bishop Wilberforce and Darwin's advocate, Huxley, debated the merits of natural selection. The victory went to Huxley because he chose to argue from knowledge in his area of expertise. Wilberforce, in venturing beyond the bounds of his understanding, made a monkey of himself.

The issues that biology teachers face today are complex and alluring. We must not be drawn to that in which we can only flounder. Science provides the factual information upon which responsible decisions can be made. If we can enhance our students' understanding of science, then we will contribute significantly to the resolution of the social and political problems that emanate from scientific discoveries.

Roy K. Baer

Omenn has effectively outlined areas of biological concern including research, innovations, international aspects, and regulation. Because the public is being made aware of these issues through the media, much controversy has developed surrounding the social implications thereof.

My concern as a biology teacher stems from the oftentimes obvious absence of any scientific understanding of an issue, as is reflected in the attitudes of many engaged in controversy. I agree with Omenn that it is imperative that discussions, and approaches to solutions dealing with directions for future research, international cooperation, and regulation be based upon solid data and understanding. All of this reverts back to education.

Omenn feels that the scientific learning involved is best left to the science classroom; issues and implications should stay within the realm of the social science class. My feeling is that the two cannot be separated. In today's high school not every student will be exposed to in-depth studies in both disciplines.

Currently, the two-year science requirement is included in the ninth- and tenth-grade course offerings. The majority of students do not go beyond this level of science study. My experience shows that eleventh and twelfth graders have more interest, keener insight, and much more maturity in dealing with these issues than do the younger students. I feel that the issues under consideration are, in many instances, "over the heads" of youngsters fourteen and fifteen years old.

As for dealing with the issues in the social science class, which follows usually in the eleventh grade, I hesitate to say so, but, many social studies teachers lack the necessary science background to handle the issues on that basis.

Because of the social implications upon us with recent developments in the field of biology and the biological consequences of technology, I have felt a critical need for a second-year required biology course in the high school curriculum. I spoke to this need at the Chicago NABT Convention in 1978. At that time a survey of teachers from across the country indicated most were in agreement with this position.

Betty W. Long

High school students are stimulated by opportunities to be at the forefront of basic research, and they are encouraged by the optimism that justifiably surrounds past and ongoing progress. Teachers can motivate them by introducing new knowledge as it becomes available, thus leading the way for students to search on their own. Small and larger group discussions can have as an objective the projection, by the students, of personally responsible courses of action in matters that range from local to global, and from scientific to social-humanitarian.

Hasty judgments are common among persons of all ages. Hence it becomes valuable for students to appreciate, early on, the necessity of obtaining information from reliable sources, understanding the basis on which the information is concluded as known, and projecting what is not known or might be learned in the future. The building of such attitudes is important for students who will not go on to higher education, as well as for those who will.

As indicated by Omenn, time needs to be taken for assessment of the applicability of knowledge to such problems as disease, food production, nutrition, maintenance and enhancement of health, prevention of waste, and protection of the environment.

Sister Marian Catherine McGrann

Omenn has outlined four important areas that will undoubtedly be important to research in the coming decades. While he points out, on the one hand, the "perils of wizardry," he tends to reject, on the other hand, the notion that the social and political issues of such research should be included in science curricula. He suggests, instead, that these questions are best reserved for classes in political science and social science. Ironically, these are precisely the classes that students of science do not take. Indeed, they avoid them.

It is true, as Omenn points out, that science is complicated enough, and students need to learn what science is about. But it seems to me that the solution is not to have scientists take courses in political science. Rather, it would seem more prudent to reevaluate goals and objectives in early science courses and provide a basis for the future scientist, as well as the nonscientist, to understand the limitations of science. True, "the role of science in public policy will continue to be challenging for both scientists and citizens," but the citizen needs to play an active, informed role. The scientist cannot operate in an ivory tower.

James L. Mariner

Omenn is complimentary of what he believes was former President Carter's leadership in bringing scientific expertise to bear on problems that otherwise might be debated strictly on their political merits. This view is not, of course, shared by everyone. When looking in retrospect at some of the actions taken by former President Carter, it appears that some decisions were politically motivated. It is too soon to know how the Reagan administration will fare in science and public policy. However, recommendations to cut the National Science Foundation funds and to cut expenditures for public education do not portend well for the advancement of science or for the establishment of a coherent, long-range science policy.

Omenn's concern about limited funding for research and development is a serious matter. His observation that policy formulation is international gives us insight into the growing recognition that science and technology are closely related to the larger political and economic issues of our times.

Regulation by government to protect health in the workplace and in the environment is important, but there appear to be excessive regulation and neglected enforcement in some areas. This is a highly controversial function of government, requiring close scrutiny.

I disagree with Dr. Omenn's position that social and political ideas ought to be taught in political science and social science classes. My view is that basic biological knowledge and scientific principles can and should be taught through the survival needs and interests of students. The impact science has on each of us requires that the social effects of scientific discoveries and their subsequent technologies be taught by teachers who know biology and can convey basic knowledge through examination of controversial issues in science and technology.

James A. Meyer

5

OUR GENES: VARIABLE AND VULNERABLE

Neil A. Holtzman

. . . in teaching human genetics emphasis should be placed on the richness of biological diversity, as well as on the frailty of each human being.

The consequences of underestimating human genetic variability may be dire. Social policies detrimental to individuals and to humankind may emerge if they are made without an understanding of the origin, extent, and inevitability of genetically determined variation, and without an appreciation of the interaction of genes and environment that results in disease.

Differences in genotypes contribute to the extensive and often continuous variation in measurable characteristics and responses to the environment. These genetic variations depend on mutations. Although some mutations will be advantageous or neutral, most will be harmful. Most mutations are caused by natural forces, such as cosmic rays and terrestrial sources, and are inevitable. Although mutagens made by humans currently cause only a small fraction of mutations, the proportion may be on the increase. Approximately 20 percent of the few thousand chemicals tested so far have been found to be mutagenic.

Gene frequencies are responsive to environmental pressures. Today's environment is different from that in which most human evolution occurred. Alleles that were beneficial or neutral in the past may be harmful today, but persist because selection pressures have not had time to operate.

In a democracy, decisions about policy ultimately should be made by the people. Enlightened policies will not be adopted, however, or accepted by the public unless the issues are understood. And, in teaching human genetics, emphasis should be placed on the richness of biological diversity, as well as on the frailty of each human being.

A RECENT ARTICLE in the *New York Times Magazine* claimed that "most healthy couples generally have intercourse two or three times a week" (Botwin 1979). Any who are dismayed to discover that they fall outside of what the author assumes to be the "normal" range needn't get too upset. A recent survey of 100 happily married couples showed that most couples do not, in fact, have intercourse as often as two or three times a week (fig. 1) (Frank, Anderson, and Rubinstein 1979).

My point in quoting the *Times* article is to emphasize the extent to which human variability is underestimated. The consequences of underestimating variability may be dire. In the case at hand, couples who have intercourse less than two times a week may become acutely anxious; marriages might be threatened. (In some other societies, those indulging more frequently than three times a week might begin to worry, but I doubt that would happen in ours.)

Not only do most people view normality within narrow limits, they also arbitrarily assume that "abnormals" can be separated from "normals." Yet histograms or curves with approximately the same distribution or shape as the one shown for the frequency of sexual intercourse are found for a wide range of human characteristics, including intelligence quotients and susceptibility to specific diseases. Some individuals will be misclassified whenever a separation line is drawn, no matter where along the curve.

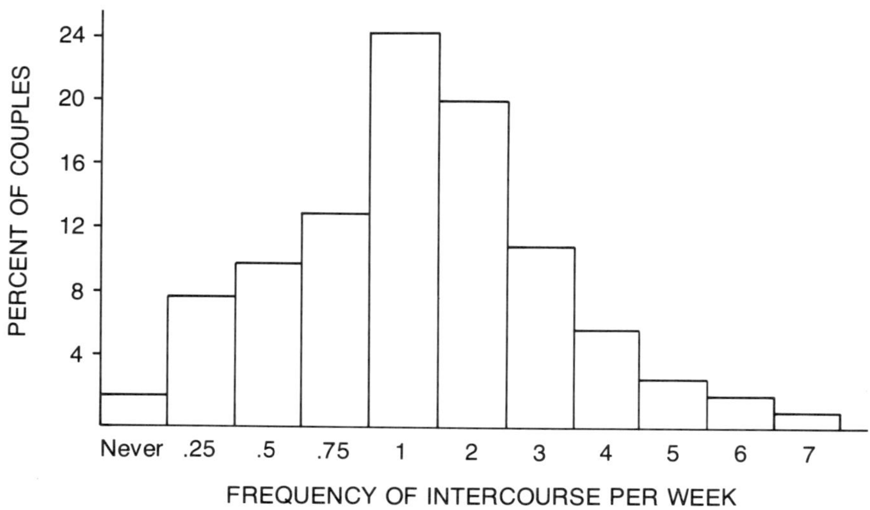

FIGURE 1. Frequency of intercourse in normal couples (mean age ± sd = 36 ± 10.5). (Estimated from Frank, Anderson, and Rubinstein, 1979.)

In some instances, the outliers may be better off than the majority. Until very recently over one-half of all males between twenty-one and sixty-four years of age smoked, and non-smokers were ridiculed. Now it turns out that the prospects for future health are better for the minority. Thus the modal range of values may not be the most healthful.

What has all this to do with human genetics? Common myths are that genetic variants can easily be distinguished from the rest of the population, and that only extreme variation can be explained genetically. It is true, for instance, that there is hardly any overlap between the blood phenylalanine levels in infants with classical phenylketonuria

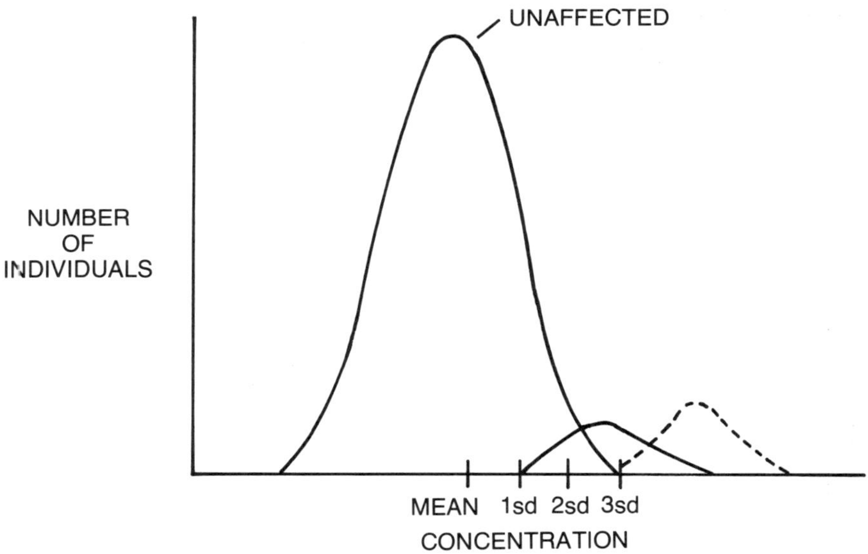

sd = standard deviation from the mean of all test results during the first week of life.

FIGURE 2. Distribution of the concentration of a substance in normal infants (large curve) and in infants affected with a hypothetical rare disorder (small curves) in which the substance is derived from the diet and cannot be metabolized. The small solid curve indicates the results in the first week of life, when intake, and accumulation of the substance, is still low. The dashed curve indicates the results in slightly older infants. Test results in unaffected subjects are much less affected by age. Although the distribution of serum phenylanines in PKU shifts to the left as infants grow older, the shift is not as marked as that shown.

(PKU) who are screened *after* the first week of life and the levels of older infants. However, most infants are screened *during* the first week of life, and then, for nutritional reasons, there is overlap (fig. 2) (Holtzman 1978). Infants with other genetically determined abnormalities of phenylalanine metabolism are even more likely to have blood phenylalanines that overlap with the distribution in normal infants. Although infants with very high levels of blood phenylalanine are more likely to become retarded than infants with less elevated levels, there is no sharp dividing line. Treatment, which is effective in preventing retardation, must be started in some infants not destined to become retarded in order to protect all of those who are at risk.

At least part of the variability among normal individuals is genetically determined. Harry Harris and his colleagues measured the activity of a red blood cell enzyme, acid phosphatase, in several hundred healthy subjects and observed a smooth, unimodal distribution (fig. 3).

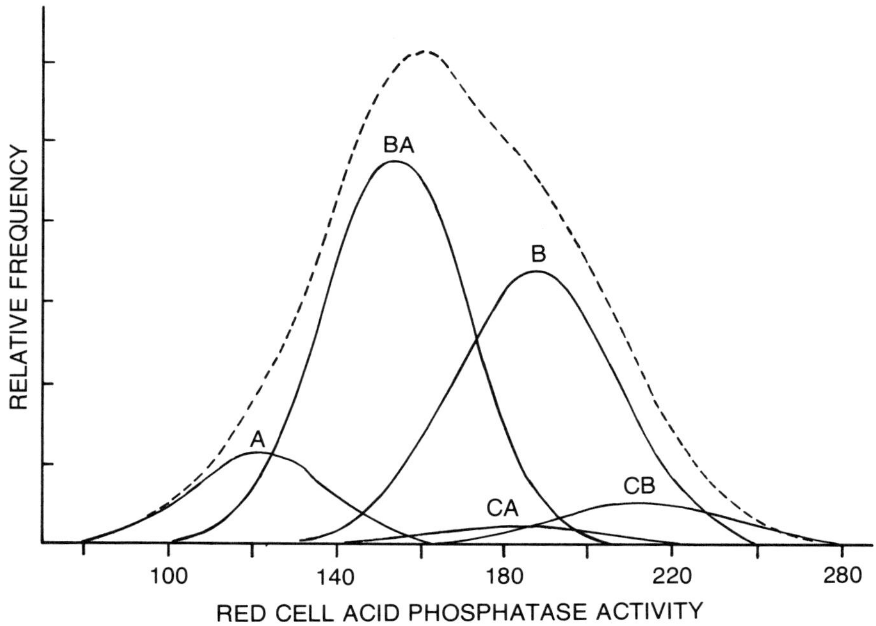

FIGURE 3. Distribution of red cell phosphatase activities in the English population. [Harris, H. 1977. *The principles of human biochemical genetics.* Amsterdam: North-Holland Publishing Company. Courtesy of Elsevier/North-Holland Biomedical Press.]

But more sensitive techniques indicated that the unimodal curve was a composite of several genetically distinct subpopulations, each with a different range of activity (Harris 1970).

For many other traits, such as the frequency of sexual activity or smoking, the variability stems largely from differences in external factors, including other people, that impinge on individuals. If those factors could be held constant, the response would still vary from individual to individual. And the basis of that biologic variation, in ways that are often complex and still poorly understood, must be genetic.

The same problems that result from underestimating variability in general come back to haunt us when genetic variability is underestimated. Without understanding the origin, extent, and inevitability of genetically determined variation, and without appreciating the interaction of genes and environment that results in disease, policies detrimental to individuals and to humankind may emerge. Let me now discuss each of these areas in turn.

Origin, Extent, and Inevitability of Genetic Variation

Genetic variants arise as a result of mutation. The simplest mutation involves the change of one nucleotide base along a sequence of DNA to one of three others. If the altered DNA is part of a gene that codes for a protein, or more specifically for a polypeptide chain, then as a result of the mutation the structure of that protein might be altered. How frequently might we *expect* to find evidence of mutation and how often are mutants *actually* found?

Consider a polypeptide chain consisting of 150 amino acids (table 1). Each amino acid is coded for by a sequence of three nucleotide bases, so that 450 are needed to code for the polypeptide. If any one of those bases mutated to any other, then 1,350 mutant alleles would arise. Approximately 23 percent of all possible nucleotide base changes would result in triplets that coded for the same amino acid as in the wild type. An additional 3.7 percent of base changes would prematurely terminate chain synthesis and would seldom be compatible with life. That leaves us with the possibility that as many as 984 variants of the polypeptide chain, each differing from the other by a single amino acid, might exist in the population if all possible mutations occurred.

Is there any evidence on how many actually do exist? Over the past

TABLE 1. The Extent Of Genetic Diversity

Polypeptide length	150 amino acids		
Minimum DNA sequence	450 bases		
Maximum number of mutant alleles arising by single base changes 3 × 450			1350
Less:			
23.4% for synonymous mutants due to code degeneracy		316	
3.7% for chain terminating mutants		50	
Number of polypeptide chains differing by a single amino acid substitution			984
Number detectable by electrophoresis			328
Number observed			
β chain, hemoglobin, 146 amino acids (33% of theoretical)			104

twenty years, samples of hemoglobins from thousands of individuals have been subjected to electrophoresis. This technique is capable of detecting amino acid alterations that change the charge of the hemoglobin molecule. Such changes comprise about one-third of all possible mutations. Approximately one-third of all of the possible variants in which a charge change is possible have been found for the beta-hemoglobin chain. There is no evidence that the beta-globin gene is more likely to be mutated than any other.

Additional evidence for the extent of genetic variation comes from work of Harris and his colleagues. They determined the electrophoretic mobility of 43 different enzymes in an average of 2,500 individuals for each enzyme and found a total of 56 variants. As with most of the hemoglobin variants, many of these were rare, occurring less than once in 1,000 individuals. From the data, Harris concluded that each of us stands a chance of one to two in 1,000 of carrying a rare variant at a specific gene locus. If there are approximately 30,000 gene loci coding for protein structure, then each of us is likely to be carrying at least 30 rare alleles (Harris 1978).

In addition to the rare variant alleles, there are common protein variants each of which occurs in at least 2 percent of the population.

The genes for these common variants are called polymorphic. Blood group substances, red cell acid phosphatase, and sickle hemoglobin are the products of polymorphic genes. About one-third of all gene loci for enzyme structure are polymorphic, and at each gene locus, each of us has, on the average, about a 20 percent chance of being heterozygotes (Harris 1978). This is only an average estimate, and at some loci the chance for heterozygosity is much higher. For instance, at any one of the four loci for the histocompatibility (HLA) genes, we each have an 80 percent chance of being heterozygotes. There are so many different alleles at each of these four polymorphic loci that 300 million different combinations are possible (Bodmer and Bodmer 1979). Thus with the exception of identical twins, the chance that any two of us carries the same set of rare or polymorphic alleles is vanishingly small.

Family studies indicating that most mutations go back several generations raise the questions about the evolutionary significance of mutant genes. It is doubtful that any of them diminish reproductive fitness, at least when they are present in single dose; if so they would not have persisted. Mutant alleles that cause serious or lethal disease when present in double, or homozygous, dose may attain high frequencies because in heterozygotes they, or genes closely linked to them, confer a reproductive advantage, or because of inbreeding. Approximately one out of twenty-five people carries the gene for cystic fibrosis and one out of sixty for PKU. No heterozygote advantage for either has yet been convincingly demonstrated.

Alleles that are neutral, that is, they convey neither reproductive advantage or disadvantage, may also attain high frequency by random drift, linkage to advantageous genes, or in-breeding. To say that a mutant gene is "neutral" merely means that the gene does not alter reproductive fitness; it does not mean that the gene is without any effect. In more than half of the known polymorphisms, significant differences in enzyme activity exist between the enzyme phenotypes that comprise the polymorphism (Harris 1978). Such differences make a substantial contribution to biologic individuality.

Variants that improve reproductive fitness eventually will be established as the most frequent genes at their respective loci, provided they are not harmful when present in double (homozygous) dose.

It may appear from this discussion that mutation is relatively benign. Remember, however (table 1), that two-thirds of all beta chain globin mutations that would be detectable by electrophoresis have not been found. A high proportion of them may not be compatible with life. Mutations in DNA sequences other than those coding for the amino acids of the protein itself, such as in the leading or intervening DNA segments of the gene for the protein, might also prevent synthesis of the protein and be incompatible with life. On balance, if the mutation rate were to increase, the reproductive capabilities of the population as a whole would be impaired. There is also considerable evidence that mutations in somatic cells are early events in the pathogenesis of cancer (Ames 1979). In addition, the appearances of fresh mutations that are harmful only in the homozygote would impose an increased burden of recessive disease on future generations. Thus, increases in the mutation rate should be viewed with alarm.

Most mutations are caused by natural forces, such as cosmic rays and terrestrial sources, and are inevitable. Although mutagens made by humans currently cause only a small fraction of mutations, the proportion may be on the increase. Almost 1,000 new chemicals are introduced in commercial quantities each year (Ames 1979), and approximately 20 percent of the few thousand chemicals that have been tested thus far in bacteria or animals have been found to be mutagenic (McMahon, Cline, and Thompson 1979; Maugh 1978).

Currently, an average person receives almost half as much radiation from artificial as from natural sources. The National Academy of Sciences Committee on the Biological Effects of Ionizing Radiation estimates that an additional 60 to 1,100 human hereditary disorders will result per million live-born offspring whose parents receive one rem of radiation above background before conception (NAS 1979). (This is equivalent to about 20 chest x-rays.) A meltdown at Three Mile Island would have caused at least 25 times as many disorders, based on a linear extrapolation (Holtzman, in press).

Genetic-Environmental Interactions

Proliferation of new chemicals and of artificial sources of radiation are symbolic of general increases of productivity and technology that have transformed the lives of people in developed countries over the last 100 years. By and large the results have been beneficial to human health as evidenced by a decline in age-adjusted mortality rates (fig. 4).

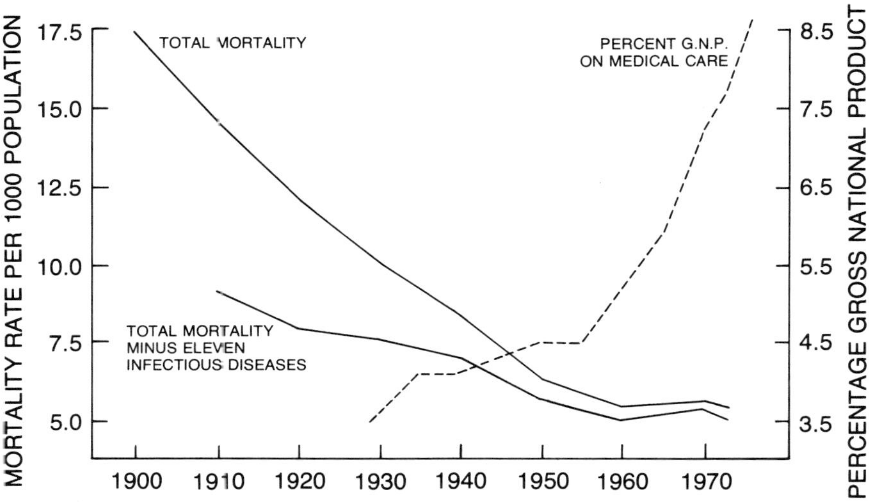

FIGURE 4. Age and sex adjusted mortality rates for the United States in 1900-1973, including and excluding 11 major infectious diseases, contrasted with the proportion of the gross national product expended on medical care. [McKinlay, J. B., and McKinlay, S. M. 1977. The questionable contribution of medical measures to the decline of mortality in the United States in the twentieth century. *Milbank Memorial Fund Quarterly/Health and Society* 55:405-428. Courtesy of Milbank Memorial Fund.]

It should be noted that improvements in nutrition, sanitation, hygiene, clothing, and housing have contributed far more to the decline than has medical care (McKinlay and McKinlay 1977).

The extraordinary change in our environment has, however, caught us off-guard genetically (fig. 5). The current human gene pool evolved under environments remarkably different from today's. Most evolution occurred in the prolonged period when humans were hunter-gatherers. Does that make a difference? Is there evidence that the frequency of certain alleles depends on environmental conditions? I will cite three supporting examples. The first is the sickle-cell gene which is found with higher frequency in populations inhabiting areas where *Plasmodium falciparum* malaria is endemic than elsewhere; individuals heterozygous for the sickle gene have some protection against this parasite. Second is the gene for the persistence of intestinal lactase into adulthood. Adults in whom the gene persists can utilize milk and other dairy products as nutrients more efficiently than those in whom it

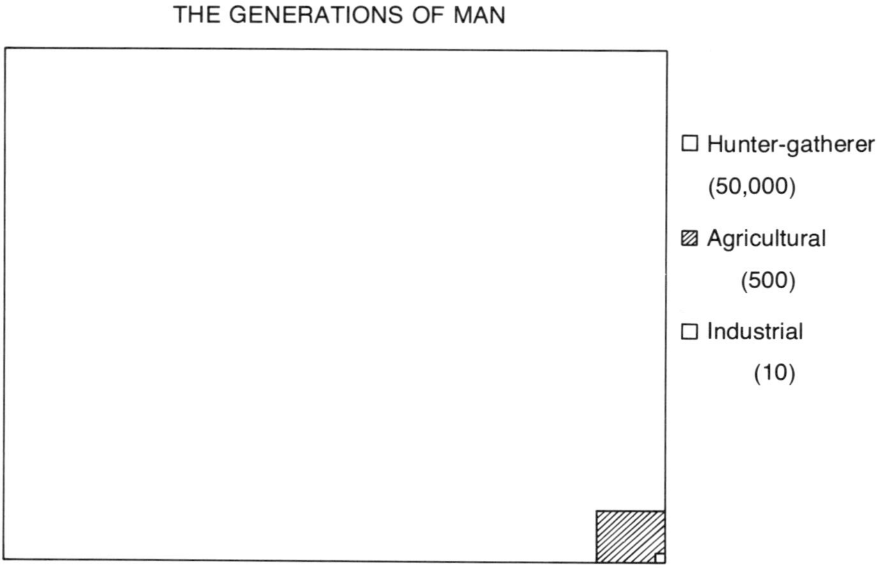

FIGURE 5. A scale drawing of the relative amount of time (in generations) that humans have spent as hunter-gatherers, in agricultural society, and in industrial society.

disappears in early childhood. The gene or genes for persistence of lactase occur with higher frequency in population groups long dependent on dairy products. Third is the gene for the Duffy blood group. This allele was recently discovered to be a receptor for *P. vivax* malaria. Today, this blood group is extremely rare in people from areas in which the malaria was endemic, suggesting that those without the allele gained a selective advantage (Miller, Mason, and Dvorak 1975).

Thus gene frequencies are responsive to environmental pressures. Today's environment is new, however, and genes that are harmful in it, but not in earlier environments, have not yet been selected against. As a result, certain diseases with strong genetic components have become fairly common.

James Neel has suggested that the genes for the adult form of diabetes may have been advantageous in the prolonged period of the hunter-gatherer when food was often scarce (Neel 1962). When food became persistently abundant the diabetes genes became disadvantageous. Although resistance to insulin eventually develops in those with the genes for diabetes, there may initially be an increased concentration of insulin. This could stimulate intake when food is abundant and improve glucose retention so that a greater reserve is present to cope with periods of food scarcity. The diabetic genotype might also result in slower oxidation of glucose and greater resistance to starvation. According to the hypothesis, disease develops when food is persistently abundant and the action of the diabetes genes results in obesity, the development of insulin resistance, and hyperglycemia. The observations that symptomatic diabetes mellitus in western populations declines during modern periods of food scarcity and that it is more likely to develop in individuals from nutritionally marginal cultures who migrate to affluent ones than in those who remain behind (Renold, *et al.*, 1978), support the hypothesis that a large, steady supply of food is needed before the harmful aspects of the diabetic genotype become manifest. Inbred mice that are heterozygous for a single diabetes gene also resist starvation better than normal homozygotes (Coleman 1979).

In parts of the world where malaria has been reduced, the net effect of the sickle-cell allele is harmful. Too few generations have elapsed since malaria ceased to be a problem to people in those areas, however, to reduce the frequency of the allele. Glucose-6-phosphate dehydrogenase (G6PD) deficiency, an X-linked condition (for which many alleles exist) is also found in high frequency in individuals whose ancestors came from areas of the world in which malaria was endemic. It probably had a protective effect, but today its predominant effect is a predisposition to hemolytic anemia on contact with any one of a number of oxidative drugs or chemicals. As with the sickle gene, an insufficient number of generations have passed to lower the frequency of the G6PD deficiency alleles.

Certain HLA alleles that occur in a high proportion of the population have been found to increase the risk of developing certain disease (table 2) (Bodmer and Bodmer 1979). However, even for ankylosing spondylitis, where the B27 allele increases the risk of developing the condition over 100 times, and where 90 percent of those with the condition have the B27 allele, only a small proportion of people who have the B27 allele will ever develop the condition. Other factors as yet unknown must be present. The high frequency HLA alleles, or alleles closely linked to them, must have been advantageous in the past, and some may still be advantageous even if they predispose to disease

TABLE 2. Relative Risk Of Disease In Individuals With Certain Histocompatibility (HLA) Groups

HLA Group	*Disease*	*Relative Risk*
D(R)w3, B8	Celiac disease	64.5, 8.1
	Chronic active hepatitis	3.4, 6.1
	Myasthenia gravis	2.3, 3.1
	Graves' disease	5.1, 3.6
D(R)w3, B8, D(R)w4	Juvenile-onset diabetes	1.8, 2.5, 3.6
D(R)w4	Rheumatoid arthritis	7.2
	Myasthenia gravis: Japanese	2.7
	Juvenile-onset diabetes: Japanese	3.4
D(R)w2	Multiple sclerosis	2.5
B27	Ankylosing spondylitis	103.5
	Reiter's disease	40.4
A3	Hemochromatosis	9.7
Cw6	Psoriasis: Caucasoids	3.3
	Japanese	15.0

Source: Bodmer, W. F., and Bodmer, J. G.: Evolution and function of the HLA system. *British Medical Bulletin* 34:309-316, 1978.

under some circumstances. A gene may have several effects, some helpful and some harmful.

Large segments of the population may be genetically susceptible to toxic reactions from pharmacological agents or chemicals that have only been introduced recently; there has been too little time, and perhaps, too little exposure, to select for protective genotypes. Approximately half of the population are homozygous for an allele for an enzyme that slowly acetylates compounds with reactive nitrogen groups. Individuals who are homozygous or heterozygous for the other common allele can rapidly acetylate these compounds. As shown in table 3, individuals receiving procainamide, a drug widely used to prevent and treat cardiac arrhythmias in heart patients, are more likely to develop antinuclear antibodies if they are slow acetylators. They also develop the antibodies more quickly and at lower doses (Woosley, *et al.,* 1978). Antinuclear antibodies frequently, but not always, herald the appear-

TABLE 3. Effect Of Acetylator Phenotype And Duration Of Procainamide Therapy On Appearance Of Antinuclear Antibodies (ANA)

	DURATION OF THERAPY			
	Three months		*Five months*	
Acetylator type	*ANA present*	*ANA absent*	*ANA present*	*ANA absent*
Slow	7	4	11	0
Rapid	0	9	3	6

Source: Woosley, R. L., *et al.,* Effect of acetylator phenotype on the rate at which procainamide induces antinuclear antibodies and the lupus syndrome. *New England Journal of Medicine* 298:1157-1159, 1978.

ance of a serious connective tissue disorder, systemic lupus. The risk of developing lupus even without contact with a specific drug, is higher for slow than for rapid acetylators. This finding has led to speculation that exposure to nitrogen-containing azo dyes used in food processing and hair sprays may account for disease in genetically susceptible slow acetylators (Drayer and Reidenberg 1977).

Other studies indicate that not all acetylators receiving procainamide develop antinuclear antibodies or lupus, and, as shown in table 3, some rapid acetylators do develop them. Clearly, there are other factors that operate.

Improvements in life expectancy have made us more aware of the presence of genes whose harmful effects become apparent in later life, after reproduction is completed. Such genes will not be affected by selective pressures. As the proportion of the population that lives beyond the age of reproduction increases, the disease-promoting features of these genes will become more evident. Coronary heart disease, lung cancer, and emphysema do not usually occur until after the peak reproductive years. There is now evidence for a genetic contribution to the development of these conditions in some individuals (Goldstein, *et al.*, 1973); Tokuhata 1964; Cohen, *et al.,* 1977). Interactions with environmental factors such as high fat diets, stress, and cigarettes are prerequisites for the development of the respective diseases.

Let me summarize the points that I have been trying to make:

1. At many gene loci more than one allele exists.

2. Differences in genotypes contribute to the extensive and often continuous variation in measurable characteristics and responses to the environment.

3. Genetic variation depends on mutation. Although some mutations will be advantageous or neutral, most will be harmful. Consequently, increases in the mutation rate will be harmful.

4. Today's environment is different from that in which most of human evolution occurred. Alleles that attained high frequency because they were beneficial (either alone or in combination with other genes) or neutral in earlier environments may be harmful today.

5. The human species has not had time to develop, through mutation and selection, alleles that will protect most of its members against harmful agents that have been introduced into the environment only recently.

6. The relatively high frequency of alleles whose deleterious effect only becomes evident late in life has become apparent as mortality rates in the young and middle-aged have declined.

7. Alleles that are harmful in single dose (dominants) are not likely to occur with high frequency in the population unless they were advantageous in the past (*e.g.*, G6PD deficiency) or unless their harmful effects appear after the reproductive years (*e.g.*, coronary heart disease).

8. Most common genetic diseases involve several gene loci and an interaction between genetic and environmental factors.

Policy Implications

I would like to conclude by considering the policy and educational implications of these points. My concern here is not with the detection and management of rare, well-defined genetic conditions like PKU, or with screening for carriers of Tay-Sachs disease or sickle-cell anemia, all of which are important. Rather, I am concerned about appropriate ways of dealing with disorders in which both environmental and genetic factors play a prominent role. Let me first state the policy options in polar terms: We can continue to allow the introduction and use of potentially dangerous agents into the environment and seek to identify those with genetic predispositions and selectively reduce their exposure. Or, at the other extreme, we can ignore genetic variation in susceptibility and attempt to reduce the presence of deleterious agents in the environment

Clearly, neither of these policies will satisfy all interests, and neither will be optimal in most situations. Let me describe some criteria that should be applied in considering policies—first for specific environmental agents, and then for specific genetic susceptibilities.

Criterion 1. How essential to human need is the potentially dangerous agent? Manufacturers are much more likely to believe in the benefits of their products, be they cigarettes, food coloring, sweeteners, spray deodorants, anti-cold medicines, tranquilizers, or a variety of other substances, than a careful evaluation of health needs would indicate. Unfortunately, vast sums are spent (far more than for education) persuading Americans that they cannot do without these items.

Criterion 2. How great is the risk? Agents that cause widespread harm with little delay are certainly more dangerous than those that affect a small number of people after many years. Americans, as cigarette smoking indicates, are willing to gamble with a distant catastrophe in order to obtain immediate satisfaction. As should be evident from this paper, for many substances there may be a continuous gradation of risk. Also, a single agent, like cigarette smoke, will cause different kinds of harm. Identifying those with a predisposition to one harmful effect will not protect those who have a predisposition on a different genetic basis to another toxic effect. In some instances, consumers may not be at considerable risk, but some workers involved in the manufacturing process may be susceptible on a genetic basis to harm from exposure. In both cases, the ratio of benefits to risks must be assessed.

Criterion 3. How practical is curtailment of use? It may be easier to control the production and dissemination of substances that are still in developmental stages than to restrict those already in use. The failure of Prohibition is a case in point, although restrictions on smoking, particularly in public places, have met with greater success.

Turning now to criteria for policies regarding detection of susceptibilities:

Criterion 1. Are there reliable and valid methods for detecting genetic susceptibility? When an allele at a single locus, such as for G6PD, confers a high degree of risk, tests are more likely to identify correctly individuals at risk than when multiple genes and unknown other factors are also involved. It should be clear that the latter is frequently the case.

Criterion 2. When screening appears to be appropriate, can mass programs that retain effectiveness be launched? In this country even newborn screening for PKU still has flaws (Starfield and Holtzman 1975). Screening itself can be run for profit and quality control can suffer.

Criterion 3. How acceptable is screening for genetic susceptibility? If most individuals will not accept screening there will be little benefit. Individual preferences may then be disregarded and screening made mandatory. Such coercion is abhorrent in a democratic society, but states have mandated newborn and sickle-cell screening (NAS 1975). I recently learned that some companies, already include genetic screening in their pre-employment physicals, but I do not know whether screening is necessary for employment, whether potential employees are offered an opportunity to refuse testing, or whether the discovery of a susceptibility will preclude employment.

Criterion 4. Once individuals are determined to have a genetic susceptibility, what can be done? Denial of employment, particularly when jobs are scarce, is not a pleasant option. The courts may view it as a form of discrimination, and workers may prefer to take the chance of getting sick on the job to getting no job at all. For those conditions for which individuals must take responsibility for altering their lifestyles to reduce risk, we do not yet know whether people who are informed of their genetic susceptibility will be more likely to take action than the general population. If, for instance, it becomes possible to screen for genetic predispositions to lung cancer, will telling someone that he or she is genetically predisposed be more effective than the current general effort to get people to quit smoking? Avoidance of drugs to which patients are genetically susceptible to toxicity does seem appropriate. This will require physicians to screen their patients for such susceptibilities.

In some circumstances, then, screening for genetic predispositions to disease is an important and appropriate approach. Such screening should not, however, be adopted at the expense of reducing exposure of the entire population to hazardous environmental agents of dubious value, particularly when low doses may be harmful and when individuals cannot control their exposure. When there is no evidence for a safe threshold dose—and there is no such evidence for mutagens—reducing the amount of the agent in the environment is preferable to setting maximum permissible levels of exposure. (Setting such levels is preferable to no limits whatsoever.) In the nuclear industry, not only has the "safe" level been set ten times higher than for the general population, but workers in some jobs last only days or minutes before they receive a year's allowable exposure. This has resulted in rapid turnover of employment, and exposure of many more people to the industrial exposure limit than are employed at any one time (Gillette 1974). The pressure for jobs, as well as ignorance of the danger, compel many to take risks.

Educational Implications

In a democracy, decisions about policy should ultimately be made by the people. There must be mechanisms by which the public can influence policy. In Maryland, the legislature has established a Commission on Hereditary Disorders—a body in which consumers are in the majority—to establish policies related to genetic disorders (Holtzman/Milunsky, in press).

Enlightened policies will not be adopted, however, or accepted by the public, unless the issues are understood. As Hickman and her colleagues (Hickman, Kennedy, and McInerney 1978) and Scriver and his colleagues (Scriver, *et al.*, 1978) have documented, students' knowledge of human genetics leaves something to be desired. Barnes and Mertens have also provided evidence that many of the examples used in the teaching of human genetics are inappropriate; some of them are not even inherited traits (Barnes and Mertens 1976). Hurd has found great differences in the quantity and quality of material on human genetics in biology texts (Hurd 1978).

Instead of emphasizing the multiplicity of genotypes and continuous variation, many curricula emphasize discrete and usually rare traits. That is a reasonable starting point, but if you stop there, students may perceive that those with "abnormal genes" constitute a small and distinct group of people. Current teaching does not emphasize that we are all vulnerable, each to different insults and to varying degree. Current teaching could well generate genetic arrogance on the part of those who, to all outward appearances, seem healthy.

Curricula do change. Cloning, *in vitro* fertilization, and genetic engineering have become fashionable topics. But these techniques are, and will remain, far removed from most people. Is it not more important to have young people understand the relation between their environment and their genes? It seems to me that in teaching human genetics, emphasis should be placed on the richness of biological diversity, as well as on the frailty of each human being.

References

AMES, B. N. 1979. Identifying environmental chemicals causing mutations and cancer. *Science* 204:587-593.

BARNES, P., and MERTENS, T. R. 1976. A survey and evaluation of human genetic traits used in classroom laboratory studies. *Journal of Heredity* 67:347.

BODMER, W. F., and BODMER, J. G. 1979. Evolution and function of the HLA system. *British Medical Bulletin* 34:309-316.

BOTWIN, C. 1979. Is there sex after marriage? *New York Times Magazine* September 16:108.

COHEN, B. H., BALL, W. C., BRASHEARS, S., DIAMOND, E. L., KREISS, P., LEVY, D. A., MENKES, H. A., PERMUTT, S., and TOCKMAN, M. S. 1977. Risk factors in chronic obstructive pulmonary disease (COPD). *American Journal of Epidemiology* 105:223-231.

COLEMAN, D. L. 1979. Obesity genes: Beneficial effects in heterozygous mice. *Science* 203:663-666.

DRAYER, D. E., and REIDENBERG, M. M. 1977. Clinical consequences of polymorphic acetylation of basic drugs. *Clinical Pharmacology and Therapeutics* 22:251-258.

FRANK, E., ANDERSON, C., and RUBINSTEIN, D. 1979. Frequency of sexual dysfunction in "normal" couples. *New England Journal of Medicine* 299:111-115.

GILLETTE, R. 1974. "Transient" nuclear workers: A special case for standards. *Science* 186:125-129.

GOLDSTEIN, J. L., SCHROTT, I. G., HAZZARD, W. R., BIRMAN, E. L., and MOTULSKY, A. G. 1973. Hyperlipidemia in coronary heart disease: II. Genetic analysis of lipid levels in 176 families and delineation of a new inherited disorder, combined hyperlipidemia. *Journal of Clinical Investigation* 52:1544-1568.

HARRIS, H. 1970. *The principles of human biochemical genetics.* Amsterdam: North-Holland Publishing Company.

————. 1978. Mutation and enzyme variation in human populations. *In* Walcher, D. N., Kretchner, N., and Barnett, H. L. (eds.) *Mutations: Biology and society.* New York: Masson Publishing.

HICKMAN, F. M., KENNEDY, M. H., and McINERNEY, J. D. 1978. Human genetics education: Results of BSCS needs assessment surveys. *American Biology Teacher* 40:285-308.

HOLTZMAN, N. A. 1978. *Newborn screening for genetic-metabolic diseases: Progress, principles and recommendations.* DHEW Publication (HSA) 78-5207 Washington, D.C.

———. (in press) Public interest in genetics and genetics in the public interest. *American Journal of Medical Genetics.*

———. (in press) Public participation in genetic policy-making: The Maryland Commission on Hereditary Disorders. *In* Milunsky, A. (ed.) *Genetics and the law II.* New York: Plenum.

HURD, P. D. 1978. The historical/philosophical background of education in human genetics in the United States. *BSCS Journal* 1(1):3-8.

McKINLAY, J. B., and McKINLAY, S. M. 1977. The questionable contribution of medical measures to the decline of mortality in the United States in the twentieth century. *Milbank Memorial Fund Quarterly Health and Society* 55:405-428.

McMAHON, R. E., CLINE, J. C., and THOMPSON, C. Z. 1979. Assay of 855 test chemicals in ten tester strains using a new modification of the Ames test for bacterial mutagens. *Cancer Research* 39:682-693.

MAUGH, T. H., II. 1978. Chemical carcinogens: The scientific basis for regulation. *Science* 201:1200-1205.

MILLER, L. H., MASON, S. J., and DVORAK, J. A. 1975. Erythrocyte receptors for *(Plasmodium knowlesi)* malaria: Duffy blood group determinants. *Science* 188:561-563.

NATIONAL ACADEMY OF SCIENCES. 1975. Committee for the Study of Inborn Errors of Metabolism. *Genetic screening: Programs, principles and research.* Washington, D.C.

———. 1979. Committee on the Biologic Effects of Ionizing Radiations. *The effects on populations of exposure to low levels of ionizing radiations.* Washington, D.C.

NEEL, J. V. 1962. Diabetes mellitus: A "thrifty" genotype rendered detrimental by "progress"? *American Journal of Human Genetics* 14:353-362.

RENOLD, A. E., MINTZ, D. H., MULLER, W. A., and CAHILL, G. F. 1978. Diabetes mellitus. *In* Stanburg, J. B., Wyngaarden, J. B., and Frederickson, D. S. (eds.) *The metabolic basis of inherited disease,* 4th ed. New York: McGraw-Hill.

SCRIVER, C. R., SCRIVER, D. E., CLOW, C., and SCHOK, M. 1978. The education of citizens: Human genetics. *American Biology Teacher* 40:280-284.

STARFIELD, B., and HOLTZMAN, N. A. 1975. A comparison of effectiveness of screening for phenylketonuria in the United States, United Kingdom and Ireland. *New England Journal of Medicine* 293:118-121.

TOKUHATA, G. K. 1964. Familial factors in human lung cancer and smoking. *American Journal of Public Health* 54:24-33.

WOOSLEY, R. L., DRAYER, D. E., REIDENBERG, M. M., NIES, A. S., CARR, K., and OATES, J. A. 1978. Effect of acetylator phenotype on the rate at which procainamide induces antinuclear antibodies and the lupus syndrome. *New England Journal of Medicine* 298:1157-1159.

Commentaries

Each of us is different. Because of our differences each of us is unique and is capable of living in a unique way. What better way to help a student feel good about being alive than by recognizing the individual uniqueness of biological diversity. Through a study of human genetics as continuous biological diversity, a person can develop a sense of belonging. Each person's uniqueness becomes a badge of acceptance and not a mark of difference.

Since each person occupies a unique environment, everyone has some responsibility for the condition of that environment and some control over it. Our health and well-being result from an interaction between genome and environment; and we can decide to a great degree what direction that interaction will take. Additionally, we must all bear the responsibility for the condition of our collective and corporate environment.

Human biology, then, becomes a potent tool for developing an expanding awareness of the biological forces that help to mold our social interactions. It becomes a discipline that can enrich a student's capabilities in responsible decision-making. It can help make the student want to be a part of the decision-making process. Once that is achieved, we have gone a long way toward meaningful education.

Roy K. Baer

Present high school genetics courses approach the subject, by and large, from the historic point of view. Gregor Mendel and his garden peas introduce the passing of inherited traits from one generation to another. After considering Dr. Holtzman's ideas, one questions whether or not we are "behind the times" with this approach. I feel that in teaching the mechanics of heredity, and in predicting possible patterns of inheritance, Gregor Mendel is still highly relevant.

Holtzman has pinpointed the need to educate young people as to the frailty of the human condition. I agree wholeheartedly with this point. Opportunities for stressing this vulnerability are present throughout the entire biology course—from cell to ecosystem—and need not be "saved" as one major concept to be presented within the genetics unit. This recurring theme should be emphasized at all levels of study, illustrated with concrete examples and case histories.

Within the genetics unit emphasis should be placed upon normal variation and phenomena which result in new genetic combinations. Students need to understand random distribution of chromosomes during meiosis, crossing-over, multiple alleles, multiple genes, nondisjunction, and mutations as the basis of genetic variation and genetic vulnerability.

In addition, it is necessary to develop units of study dealing with harmful substances and mutagens and carcinogens in the environment. Students have opportunities to evaluate both the needs and the risks inherent in the introduction of certain chemicals into our lives. Students, as future policy-making citizens, need to understand the dangers associated with the development and use of deleterious agents. We need not look beyond the daily newspaper for evidence of the tragedy involved in ignoring this issue.

Betty W. Long

Dr. Holtzman's allegation that genetics as taught in many biology classes today is inadequate or "trendy" seems to me to be basically correct. Some of my students *are* confused and do make statements indicating that they are given to genetic snobbery because they do not happen to belong to a "statistical" group or ethnicity. There is a need to emphasize and adequately illustrate the great extent of genetic variability in the family of humankind.

If today's students and tomorrow's citizens are going to participate effectively in the decision-making which will affect their lives and their futures, they certainly need the understanding that will come from learning about human genetic variability. Such instructional experiences should be accompanied with examples of individual vulnerability and how modern artificial selection acts on our genes.

James A. Meyer

It is extremely important for students to know that people are different from each other. Students should realize that there is a relationship between their genetic makeup and their environment and that although they cannot do much about their genotype, they play a tremendous part in the selection, creation, and maintenance of their environment.

It should be a relief to many students to know that they can be different from their peers without being ostracized. People who are different should be the rule, not the exception, and the genetically different should not be looked upon with tunnel vision or distrust. There *is* a richness in diversity. There is little progress if there is no difference among people. It is the clones who are unnatural. Sameness can lead to disaster as seen in our monocultures of corn and wheat and the never-ending battle against their parasites.

A comment commonly heard when a product has been newly found to be associated with cancer is, "What doesn't cause cancer?" Most things don't, but it is the nature of the world we live in today that artificial agents that can cause an increase in the mutation rate *are* developed, used, and discarded into our environment. Each student must personally weigh the risks versus the benefits of using a known mutagen or being in its presence, but each also needs information on which to base a decision.

Gordon E. Uno

6

HUMAN SEXUALITY AND REPRODUCTION

Elizabeth J. Roberts

Sexual learning is social learning. It begins at birth and continues throughout life.

Despite what appears to be our sophistication and preoccupation with sex in our country, millions of Americans of all ages need and want help in understanding their sexuality.

Our sexuality is part of our basic identity. It is expressed in the full range of relationships we have with others; sexuality influences—and is influenced by—interpersonal relationships, family roles, and lifestyles.

In 1978, Roberts and her colleagues completed a major study on parent-child communication about sexuality. Based on the results of that study, Roberts concluded that "children today are still learning about the birds and begonias, but not much about themselves." The vast majority of children in our country receive little or no information about erotic dimensions of sexuality from their parents or from most other authority figures in this society. They're forced to come to an understanding about a critical part of their personality without the guidance and support of their parents.

Beneath the picture of the status quo, however, there are rumblings of change. Our task for the future is to expand people's understanding of sexuality, defined most broadly. That amounts to nothing less than having us think differently about ourselves and the world in which we live.

DURING THE PAST ten years, three presidential commissions, countless organizations (both national and international), and many professionals have stressed the importance of human sexuality in our everyday lives. At the same time, they have documented the ignorance, misinformation, and mythology about sexuality that generally pervade our society. Ignorance about human sexuality contributes to poor self-image, faulty communications, uninformed decision-making, and unnecessarily stressful interpersonal relations; and the situation is self-perpetuating. A population that deals uneasily with sexual issues is not likely to guide the sexual learning of its young people in any meaningful or effective way. Today—despite our presumed sophistication and our preoccupation (or at least the media's preoccupation) with sex—millions of Americans of all ages need and want help in understanding their own sexuality.

In 1975, John D. Rockefeller III established the Project on Human Sexual Development to explore the current approaches to learning about sexuality in our society and to suggest relevant ways of improving the conditions of sexual learning. The mandate was difficult. Two of the most difficult obstacles to overcome were the widely held popular attitudes that (1) sexuality is limited to "sex" and reproduction; and (2) learning about sexuality is limited to formal, compartmentalized sex education programs.

The Meaning of Sexuality

For many if not most individuals, the term "sexuality" brings to mind a brief and limited encounter that occurs at a certain time, in a certain place, with a certain person, that uses specific parts of the body and frequently results in conception. This narrow view teaches children that sexuality is something separate from the rest of their lives; and in later life, can create a preoccupation with genital performance and sexual adequacy. It can lead to anxiety and eventually dysfunction. Further, it fosters a view of sexuality concerned primarily with reproduction. But "human sexuality is not confined to the bedroom, the nighttime, or any one part of the body." To quote the American Medical Association:

> Sexuality involves what we do, but it's also who we are. It is an identification, activity, a biological and emotional process. It is an outlook and an expression of the self. It is an important factor in every personal relationship and in every human endeavor from business to politics. (American Medical Association, Committee on Human Sexuality, *Human Sexuality,* 1972)

Understood in this larger context, sexuality involves considerably more than intercourse, reproduction, or "social hygiene." Our sexuality is part of our basic identity. It encompasses our total sense of self as male or female. It involves our attitudes about masculinity and femininity. It involves how we grow to feel about our physical selves—the limits, the joys, and the embarrassments of our bodies. It is the integration of our needs for affiliation and love and our expressions of affection and intimacy as well as our fears, fantasies, and our decisions regarding personal conduct. Human sexuality is expressed in the full range of relationships we have with all other people in our lives. It influences and is influenced by, interpersonal relationships, family roles, and social lifestyles.

Sexual Learning

In light of this broad meaning of sexuality, education about sexuality includes considerably more than a sixth grade course on the facts of life or a lecture on family planning and contraception. Sexual learning is social learning. It begins at birth and continues throughout life. Individuals do not magically become sexual at puberty or in the marriage bed. We begin learning about sexuality the day we are born. We learn from our own experiences and from the actions and reactions of others. In addition to information, we acquire values, attitudes, and behavior patterns. We learn from books, television, toys, and playtime. We learn from parents, relatives, friends, and teachers. We learn from school and work policies, as well as from the messages communicated through our major social institutions. The "silent" pediatrician communicates as much as the X-rated magazine stand.

The Project on Human Sexual Development

For five years, the Project on Human Sexual Development explored childhood learning about sexuality—the learning that occurs before children reach thirteen and society begins to treat *them* as sexual human beings. We've paid particular attention to parents, television, and community service providers and the role these three play in informing children about sexuality. In this discussion, I will focus my remarks on our findings concerning parent-child communication (verbal and nonverbal) about sexuality.

Our study was part of a community program established in Cleveland to examine the politics of sex education at the community level. We interviewed over 1,400 mothers and fathers of three- to eleven-year-old children and asked about a wide range of attitudes, values, and behaviors. In the field of sexuality, too often, studies have tended to explore only one aspect of sexuality or target one particular population. We were trying for a much broader understanding. What is the context in which people understand sexuality and what do they try to communicate to their children? We asked parents what they learned growing up; about their own child-rearing practices today, their values, and attitudes about family life and lifestyles, gender roles, their expressions of affection, and about erotic conduct and its social consequences. They talked to us about their own patterns of sexual behavior, what they have said to their children, and what they hope their children will know and believe by the time they are adolescents.

The parents were not a homogeneous group. They ranged in age from sixteen to sixty. Some were professionals; some were blue collar workers. Among them were black and white, Protestants, Catholics, Jews, and Atheists. All had attended some school (and we later found that parental educational level was related to differences in parental attitudes). Most had attended high school. Others had gone only as far as junior high; a few had graduated from college. Most of the mothers and fathers we interviewed were married. A substantial portion of them, however, were divorced, widowed, or separated; a small number of the mothers had never married. In short they were not very different from the general citizenry of the United States today. In many ways, the parents' attitudes were as diverse as their lifestyles.

Results of the Interviews

As previously mentioned, sexuality encompasses a broad range of values and attitudes. To understand how a parent feels about one dimension of sexuality does not necessarily make it easy to predict how he or she feels about another. It is hard to draw typologies of parents, liberal versus traditional. For example, many parents who had liberal attitudes concerning gender role had traditional or conservative views on erotic behavior. The major differences in attitudes were most often related to the educational attainment or the gender of the respondent.

Most other background variables (with the exception of whether or not a woman participated in the paid labor force) made very little difference in parental attitudes.

Most parents, regardless of education, religion, age, or marital status do not talk to their children about sexuality. Today's parents are struggling to reconcile new realities and new attitudes with the messages they received while growing up. They are clearly struggling to reconcile their behavior with their beliefs and their hopes for their children. As a result, most parents are confused, uncertain, and uneducated about sexuality. For the most part, parents (from all social classes and all educational backgrounds) retreat into silence and do not discuss sexual issues with their children. When asked why, the great majority of them say they are "waiting for their child to ask." But while they wait, their child is learning not to ask. Opportunities that are lost when a child is five or eight or ten may be lost forever to most parents. If one has not talked to a child about sexuality before adolescence it is almost impossible to do so after. The issues are likely to have become too fraught with anxiety and tension between parent and child.

A number of other unfortunate consequences result from this pattern of "waiting." For one thing, it puts the chief responsibility for initiating any conversation about sexuality on the child, so that information is limited by the child's ability to ask the right question. As most of us know, frequently you have to know the answer to ask the right question. Thus, most areas of sexuality—and many areas of human life—go undiscussed. Not surprisingly, if a question is asked, it is usually asked by a younger child who has not yet learned the taboos. Usually the question is simple and basic. Where do babies come from? Why are boys and girls different? Why are you and mommy getting a divorce?

Pregnancy and birth, physical differences between men and women, and marriage and divorce are the three most frequently talked-about topics within most homes today. However, even on these topics, parents' answers are brief, simplistic, and often intended to discourage further questions. Let's look more closely at what parents reported.

Three-quarters of the mothers and half of the fathers said they had talked with their child at least once about pregnancy and birth. At first glance, those data were encouraging. However, most of the parents who reported that they had talked about this topic said they did so in terms of animal life. Children today are still learning about the birds and the begonias. Only one-third of the parents who ever mentioned this topic did so in human terms, and only 15 percent of the mothers and 8 percent of the fathers ever mentioned intercourse. Sixty percent of the parents in our study said they had discussed physical differences; 40 percent said the topic had not come up in their family at all. Of those 60 percent who had discussed physical differences, most told their children that boys and girls should use different bathrooms or that God made boys and girls different. Only a minority said that boys have a penis and girls, a vagina. (Not one parent in 1,400 reported ever mentioning the clitoris as a part of female anatomy.)

Two other topics related to physical development that are rarely discussed in the home are menstruation and wet dreams. Less than 1 percent of mothers and 2 percent of fathers had mentioned wet dreams to a son or daughter. Less than 5 percent of the fathers had ever discussed menstruation with their child. Mothers were, not surprisingly, more likely than fathers to talk about menstruation. Nevertheless, 40 percent of the girls between the age of nine and eleven (the oldest child age group in our sample) have *never* had a parent mention menstruation to them. From follow-up discussions with the parents, we learned that reluctance to discuss facts of life was related to the lack of discussion most of the parents themselves had in their families when they were growing up. A number of them reported their own childhood experiences with sadness and even anger. However, most said they do not know how or when to approach these issues today. Fearful of saying something wrong, they say nothing at all.

Masturbation is another difficult topic for most parents—one that very few discuss and most are confused about. Approximately 85 percent of the interviewed parents said they believe that most children masturbate. Only slightly more than 60 percent approve; however, approximately 40 percent of both mothers and fathers reported that genital play among young children was not all right. Most parents reported being comfortable with genital play or self-exploration when their child was an infant. But at some point, this activity crosses some amorphous gray line and changes from genital play to "masturbation" (with its erotic overtones). When that happens, it causes problems for parents, especially for parents of daughters. This is one of a number of areas where the double standard is still alive and well. Most parents are prepared to tolerate their son's masturbation or future premarital sex but not their daughter's. This certainly influences the different learning that girls and boys receive about their body and how they experience their body as they move into adolescence. The majority of parents who disapprove of masturbation do so primarily for religious reasons or for fear of physical or mental harm. Even those who approve or say they approve frequently give their child ambiguous messages. One mother in our study who said she approved of masturbation, told her seven-year-old daughter: "As long as you do it in bed, don't overdo it, and don't get sick from it, it's fine."

It seems then, while parents want their children to understand and feel good about their bodies, the messages many children receive today may have other effects. Do parents ignore a child's need for assurance about body changes? Do they structure family life in such a way that questions cannot be asked easily? Do they communicate that touching or exploring the body is harmful? If they do, children will learn that ignorance is bliss, that bodies are a source of shame, and that physical development is not an appropriate topic of conversation. This kind of learning may make simple health precautions in adolescence and adult life, such as a breast self-exam, a check-up for VD, using a condom or a diaphragm, unnecessarily laden with anxiety. Children may grow up making decisions based on misinformation and may develop a fear of communicating emotional and physical needs. If this is what children are learning, all the information in the world may not make it easy for a girl to use a contraceptive if she's been taught that to touch herself is wrong and to talk about her physical needs and feelings is taboo.

In expanding our notion of sexuality, we do not want to ignore sexual conduct and erotic behavior. The vast majority of parents had never mentioned erotic aspects of sexual activity at any time to their child. While 75 percent of the mothers had talked about pregnancy and birth, only 15 percent of the mothers and 8 percent of the fathers had ever mentioned intercourse. Even fewer had discussed premarital sexual activity and only 6 percent had ever talked about venereal disease. And, even though 85 percent of the mothers and fathers we talked with said they believed that contraceptive information should be made available to preteenagers, less than 8 percent of the mothers and less than 2 percent of the fathers had ever mentioned birth control to their own preadolescent child. The vast majority of children in our country receive little or no information about erotic dimensions of sexuality from their parents or from any other authority figures in this society. They're left alone to make sense of isolated, random, and disconnected bits of information. They're forced to come to understand a critical part of their personality without help of parents' values, guidance, or information.

In addition to the limitations on the information that is exchanged, waiting until the child asks questions about sexuality has a number of other consequences, including the almost total absence of the father from the process of communication. If a child asks a question, he or she is most likely to go to the mother. This is true whether the mother works outside the home or not. Parents (both mothers and fathers) reported that they felt that fathers had limited time to spend with children. The father was usually seen as the family disciplinarian—perceived by many sons and daughters as less emotionally expressive, less accessible, or less interested in these issues than is the mother. The whole notion of "the father-son talk" in our society seems to be largely a myth. Boys, if they ask questions at all, take them to their mothers; and as they get older, boys are likely to have already learned that men are supposed to know answers without having to ask questions. Thus, boys stop asking questions altogether, relying more and more on the in peer group as a source of (mis)information. One exception to the lack of father participation was in those families where fathers shared more frequently in child care and house care. In those homes the child was as likely to ask questions of the father as of the mother. (Unfortunately, that occurred in a very small percentage of the homes in our sample.)

Finally, if parents do respond to their child, they are likely to feel that the topic is finished. It is what a friend and colleague has dubbed "the inoculation theory of sex education": if they ask, give it to them once, and you will never have to go back and talk about it again. Many of our school programs are structured the same way. We don't realize that as children grow, as their bodies change, as their intellectual abilities change, as their social context changes, they need "boosters." They want you to talk about the same issues again.

It is not only parent-child conversation that informs young people about sexuality. Children learn from observing. They watch their parents express affection or not. They learn from the roles they see. On the face of it, the roles and organization of family life among the parents we talked with are very traditional, looking much the same now as they did a generation ago. Mothers, even if they are employed outside the home, still have almost full responsibility for care of home and children. Fathers are less emotionally expressive, less inclined toward affection, and less likely to show vulnerability.

There are important exceptions to this pattern, however, and underlying this traditional picture are strong rumblings of change. While both mothers and fathers said that the mother has all the responsibility for care of the house and child, 50 percent reported they do *not* think it should be this way. Half also said they do not believe that men and women should have different roles in life.

In our sample, 40 percent of the women worked full-time in the paid labor force. The impact of labor force participation on the attitudes of women was one of our more surprising findings. At times, this life experience seemed more influential in predicting attitudes than education, income, or all the socioeconomic status indicators typically used to differentiate between populations. For example, a junior high school-educated woman who worked on the assembly line was more likely to have what we might call liberal or progressive attitudes about many dimensions of sexuality than a college-educated woman who was full-time in the home. For women, labor force participation was strongly related to more liberal attitudes. It did not, however, change their behavior. These mothers were not talking to their children more, but their attitudes about what they would like to say (should the question arise) were more liberal than the attitudes of mothers who were full-time in the home. Changes in these women's attitudes, however, did not necessarily mean there were changes in the husbands' attitudes as well.

Women who were full-time in the paid labor force were also more likely to be dissatisfied with the current organization of their family life. Often working women want their husbands to take on more family responsibility. However, because many of these tasks are still identified as "women's work" in most segments of our society, asking men to assume responsibility for these tasks raises many issues about femininity and masculinity.

Parents, especially fathers, are uncertain about other changes occurring in definitions of masculinity and femininity. Over half of the parents said they have cautioned their sons to "act like little men" and not sissies, and they have talked to their daughters about not being tomboys. Fathers are especially likely to do this. The timing is different for boys and girls; however, for the girl, it begins somewhere in preadolescence. For the boy, the concern arises much earlier. Before beginning school, a boy is cautioned not to cry—or otherwise behave—"like a girl."

Many parents, particularly fathers, have trouble finding ways of modeling the behavior they say they would like to pass on to their children. Ninety-five percent of the parents we talked with said that they would like their sons and daughters to believe that it is all right for men to cry. Yet the vast majority of fathers seldom or never cried under any circumstances, especially not in front of their children. Fathers say that touching, holding, and showing physical affection are things they learned not to do while they were growing up and still find difficult to do today.

Still another area of parental uncertainty about sexual values can be seen in their attitudes toward marriage and children. The parents with whom we spoke almost unanimously agreed that their greatest source of satisfaction in life was marriage and children. What was surprising was that 60 percent of the parents we talked with felt their own children need not get married or have children in the future. Parents' own commitment to a lifestyle today is not, then, translating automatically into certainty about the same values for their children.

On the one hand, we see that much remains the same. On the other hand, we see strong indications of transition and change. For many parents, these changes have called into question their fundamental attitudes and values. They are not sure how useful their values are. They are not sure how accurate their information is. Rather than do something wrong, they retreat into silence and do nothing at all.

Teenagers and Sexuality

To speculate about the potential influence all this has on adolescents, I'd like to offer a hypothetical example. Let us imagine two sixteen-year-olds (a male and a female) involved in the decision of whether or not to engage in sexual intercourse. Let's look at the messages they have each received about sexuality that will influence this decision.

Perhaps our young man was raised to believe, as part of his gender role, that growing up means growing away from others: to be independent, to know answers without having to ask questions. Perhaps he has had little training in expressing his feelings and his fears. He may have learned that his body is something he must perform with to prove himself and his masculinity. Perhaps he has been encouraged by peers, teachers, and parents to be goal-oriented—to succeed at all cost, even in the bedroom. Having been taught to identify masculinity with being cool or scoring, having been taught that wanting to be touched or held or comforted might be perceived as weak or unmanly, our adolescent boy might be expected to decide that the best way he can express his masculinity, prove he is grown up, and receive affection is by pressuring for intercourse.

Now for our young woman. Perhaps she was raised to believe in the importance of meeting the needs of others, of being nurturing and caretaking. Perhaps billboards, television programs, parents, and friends have conveyed to her that the most important aspect of her body is its attractiveness to males. Perhaps she was raised to think that sex is something that happens *to* her, something she knows very little about, and something she should not or cannot make decisions about. Perhaps she is frightened or ambivalent about her own erotic feelings. The only information she has about male physiology is the inaccurate notion that men are possessed by an uncontrollable sex drive.

Will the conditions of this sexual learning permit these adolescents to make an informed, satisfying, and responsible decision, or does the girl end up pretending she was "swept off her feet"? Does the boy have to fear being thought less of a man if what he really wants is simply to be held? Do they have to worry what their parents, teacher, doctor, or minister will think of them if they ask about contraception? Have the ways in which they have learned about sexuality enabled them to talk openly and honestly with each other, to ask for guidance, or to trust adult advice?

This example is not really about adolescent intercourse. Rather it is about the many messages of sexual learning that influence what we do and how we feel about what we do. It is about body image. About feelings of love and intimacy. About what we think is sexy. About how we make decisions.

It seems that our task for the future is to expand people's understanding of sexuality. This is not easy to do. It means putting a new idea on the agendas of our schools, our communities, and our nation. It amounts to nothing less than having us think differently about ourselves and the world in which we live. The task is to teach an expanded notion of sexuality and sexual learning in the context of *all* the learning that gives meaning in our lives. This is our challenge for the 1980s.

Commentaries

Teaching human reproduction in biology courses seems to be challenge enough in these days of school boards ordering pages to be torn from textbooks, members of the "Moral Majority" invoking the Wrath of God upon school boards and teachers who deal in anything more than birds and begonias, and school administrators lacking the fortitude to defend teachers who risk their jobs and reputations by teaching about penises and vaginas (do we dare mention a clitoris?). Now Elizabeth Roberts actually challenges us to delve into the complete range of human sexuality in its broadest sense!

This indeed is the controversy of controversies that all biology teachers face. No greater skill or expertise is needed by any teacher than to be able to successfully introduce a unit or course in human sexuality and have it accepted by the community. Yet there is probably no greater need than this. Effective teaching about sexuality depends greatly upon the teacher. It requires a teacher who is informed, sensitive, and secure in his or her gender identity.

As Roberts reports, most parents believe that sexual training belongs within the family and have an innate distrust of teachers in this matter. She points out that sexual training is affected in the home by the behaviors and actions of the parents and by the subtleties of silence. Most parents seem to have a narrow view of sexuality as Roberts defines it and have forgotten that what we as adults *do* is always a stronger lesson than what we *say*.

The home is not adequately preparing people for the complex, changing attitudes, beliefs, and sexual behaviors of America in the eighties. Probably few biology teachers are prepared to do so. Many are not convinced that it is the thing to do! Some are not confident enough in their own lives to be effective teachers as role models for this delicate challenge. That leaves few who will have the opportunity, inclination, and skills to meet the need. Those who can and will dare to teach human sexuality in the public schools must work together to meet the challenge that Roberts presents here.

Roy K. Baer

The broad problem of expanding young people's understanding of sexuality, on the level presented by Elizabeth Roberts, seems to be that of guiding students toward the development of positive attitudes about sexuality. This is not something that can be presented as a unit within the curriculum of a given course. This development is a gradual process which might best be accomplished throughout the middle and high school years, if not throughout the entire school program.

Such is not the case in most school systems. In Jefferson County Public Schools where I teach, sex education is included as part of the health, general science, or home economics curriculum. Instruction of the subject is not mandatory and is generally left up to the discretion of the individual teacher. Written parental consent is required for student participation in a unit on sex education. Students without permission to participate spend that class time in the library.

Is it possible to develop broad programs encompassing the entire twelve-year school experience as a means of understanding human sexuality? The greatest need, as I see it, is in training teachers to deal effectively with this issue. Not every teacher is prepared to deal comfortably with sex education on all levels described by Roberts. Training programs are necessary to help teachers approach the subject in a manner appropriate to different age groups, and to be able to create an open and comfortable atmosphere for discussions. In addition, appropriate written and audiovisual materials are needed for classroom use.

Betty W. Long

Teenagers and young adults sometimes express disappointment and resentment when they come to realize that, as children, they were not adequately informed about sexuality and reproduction. They define the inadequacy in terms of basic facts as well as meanings and attitudes. Dr. Roberts' study and her application of it make it again clear that the very young child wants and needs direct and accurate information concerning terminology. Appropriate meanings and attitudes can be absorbed very early in an atmosphere of ease and trust. There can then be hope that the child will become the mature person who, in later stages of development, will be able to make informed and responsible decisions.

At hand are the difficult tasks of defining target populations for educational programs and defining and implementing the necessary programs. As Dr. Roberts indicates, "It seems that our task for the future is to expand people's understanding of sexuality." Perhaps it is well that we first reevaluate our attitudes toward ourselves as whole persons and toward others.

Sister Marian Catherine McGrann

I wish that this essay could be read by all educators in this country—including the television and movie industries, newspaper and magazine chains, schools, and any other organizations concerned with communicating ideas and information about our reproductive biology and our human-social sexuality. The current media "hype" and emphasis on "sexiness" and the contrived erotic adventures and misadventures of our time denigrate the beautiful, positive, social, tender, and loving aspects of human sexuality. Classroom teachers of "sex education" and reproduction especially should benefit from this essay, because of its up-to-date assessment of widely held views regarding sexuality and reproduction. It would augment and support the best efforts of the sensitive, responsible teachers of our future citizenry.

James A. Meyer

7

WHY SCIENCE AND VALUES IN THE CLASSROOM?

Joseph D. McInerney

Blind adherence to dogma is dangerous, whether that dogma has its roots in religion, politics, economics, or science.

Concerns about lagging American productivity and innovation have drawn attention to science and engineering education in this country. Science educators are responsible for preparing both the professional scientists who will propose solutions to contemporary problems and the citizens who will determine how those solutions are translated into public policy.

An informed citizenry familiar with the nature and capabilities and limitations of science and technology is essential to the conduct of good science and effective science policy. Public perception of science and technology is determined increasingly by the participation of scientists in the development of public policy. This creates confusion and dissonance among a public that has been conditioned to perceive science as a value-free activity.

Since science and technology will intervene more and more directly in our lives, science education materials and programs must present science as an activity that is rooted in human values. By so doing, science educators can help to develop a more realistic public understanding of the conduct and capabilities of basic and applied science.

WE ARE ALL, I suspect, monuments to C. P. Snow's *Two Cultures*. That seems a shame, given the confused and contentious times in which we live. A colleague of mine has a poster on the wall of his office that reads "None of Us is as Smart as All of Us." That is certainly true, given the range of problems we face and the complex and controversial solutions they will require.

Science educators face the dual task of both preparing the professionals who will propose possible solutions to those problems and of educating the mass of citizens who will help decide how those solutions are translated into social and public policy. To argue which is the more important task is to miss the point that good science and effective science policy require informed participation by both groups. Current concerns about lagging American productivity and innovation (Abelson 1980), highlight the importance of an enlightened citizenry that supports the activities of the research and development communities. Concern for science education has, therefore, taken on a new urgency.

While the problem has many facets, much of the present concern over science education and scientific literacy centers on the public's inability to comprehend the nature, capabilities, and limitations of basic and applied science (Handler 1980; NAEP 1979; NSF 1981). That inability, not surprisingly, results in confusion over the processes whereby potential solutions to science-based issues are developed.

There are those, for example, who maintain that those solutions are generated in an atmosphere where precious little consideration is given to the moral or ethical implications of basic research and its derivative technologies. One has the feeling on hearing that assertion that we are being swept blindly into some Orwellian future by a juggernaut of technology, immune to temperance by reason, compassion, or morality. On examination, however, that assertion simply does not hold up.

- Consider for example, the recent pronouncement by the National Conference of Catholic Bishops prohibiting tubal ligation, even in cases where the couple is at risk for producing a child with a severe genetic abnormality (Hyer 1980). The advice of the Bishop's Conference to such couples is to abstain from sex and concentrate on their spiritual lives. Beyond simply ignoring the importance to human beings of what they call "the genital expression of love," the Bishops also ignore, or are unaware of, the growing body of data that demonstrates that the birth of a defective child can have disastrous consequences for the couple, for siblings, and for the family in general (Kelly 1977; McCollum and Silverberg 1979; Levine 1979; MacIntyre 1977).
- Consider that a right-to-life group in Texas can apply so much pressure to the local Planned Parenthood affiliate that the United Way was forced to ask Planned Parenthood to sever its affiliation with the larger organization, lest all other United Way

agencies suffer substantial loss of funds (*Corpus Christi Caller* 1979).

- Consider that anti-sex education groups prevent sex education from being taught in many public schools, despite the fact that "fully one-third of all girls between 15 and 20 years of age have at least one unwanted pregnancy" (Holden 1979), despite the fact that "teenagers constitute a disproportionately high percentage of women having abortions in the second trimester" (Hollingsworth and Kreutner 1981), and despite the fact that recent polls indicate that approximately 75 percent of the adults in this country think that sex education should be taught in public schools (*Education U.S.A.* 1978).

- Consider that in 1979 and 1980 almost a dozen states introduced legislation that would mandate the teaching of creationism in the schools in those states—fifty-five years after the Scopes trial. Supporters of that legislation in Georgia, where the legislation was defeated, estimated that similar bills "would be introduced in as many as forty states if the bill passed" (Skoog 1980). [*Editors' note:* In 1981, such bills were passed in Arkansas and Louisiana.]

- Consider the 1975 moratorium on federally supported research on *in vitro* fertilization (Culliton 1978), and the ardent opposition to IVF expressed during the 1978 hearings of the National Institutes of Health's Ethics Advisory Board (U.S. DHEW 1979). To this day there is only one clinic in the United States conducting research on IVF—with private funds, and under the guidelines developed by the Ethics Advisory Board (Marsh and Self 1980).

- Consider that the Cambridge City Council ordered a halt to the construction of a new laboratory for recombinant DNA research at Harvard University. Construction was resumed only after a lengthy series of hearings by a special citizens' commission.

- Finally, consider that recent court cases seeking damages for wrongful life, pursuant to malpractice, have generally been disallowed by the courts at the appellate level (Capron 1979; U.S. DHEW 1979). Judges in those cases have stated that such decisions are more the province of theologians and philosophers than of scientists, legislators, or jurors.

It is difficult indeed to agree that science and technology dominate our society to the exclusion of values and moral principles. To be fair, there are arguments that can be made on the other side of the issue. One can list, for example, the 1973 Supreme Court abortion decision, or the June 1980 decision by the California Court of Appeals that a child with a genetic disorder can sue a negligent physician and his/her parents for wrongful life (*Science 80* 1980). But those decisions certainly do not indicate domination by science and technology when juxtaposed with the foregoing information.

What all of those issues do demonstrate is the significant disparity that exists between the rate of scientific and technologic progress and the rate at which new social, ethical, and legal systems evolve to deal with the implications of that progress. That is not to say that science and technology proceed in the absence of input from those domains. Scientific and technologic progress is, however, often exponential. Progress in the social, ethical, and legal domains is ponderously slow by comparison, linear at best, and sometimes nonexistent. During periods of retrenchment in the political climate, progress may give way totally to regressive public policy.

Where does that leave us? Surely we are committed to science and technology. We are wedded to them; there is no going back for a technological society. The answers to our problems, many of which were created by technology, will lie, as often as not in more—not less—technology.

Our philosophy and actions as science educators must reflect that reality. We must work toward the development of materials and programs that permit the public to make realistic assessments of what science and technology can and cannot accomplish, and we must do it in a manner that conveys the extent to which science is rooted in human values.

Technology as Human Activity

Technology is but a demonstration of our ability to mediate the environment. That is the source of the controversy surrounding science and societal values—that human beings can and do actively mediate and modify their environment. That is one of the characteristics that separates us from other organisms. According to C. H. Waddington

(Waddington 1960), the human ability to alter the environment has resulted in a profound change in the direction of human evolution, where emphasis has now shifted from the biological domain to the social and psychological spheres. That is not to say that for humans biological evolution has ceased, but rather that its manifestations, which are generally quite subtle, have been overshadowed by the implications of cultural evolution. (The subtleties of biological evolution are becoming more overt as we collectively modify the environment with chemical and other types of mutagens, carcinogens, and teratogens. The human gene pool, which has evolved in a rather pristine environment over millennia, is now being subjected to thousands of potentially harmful agents. Biological evolution does not proceed rapidly enough to adapt appropriately. See Holtzman, Chapter 5 of this volume.)

Let me return, however, to Waddington's thesis and its implications. It has been said, for example, that the shift in direction he identifies may contribute to our disregard for the environment by deflecting our attention from simple biological imperatives, such as the natural laws of the biosphere and the interdependence of all organisms (Fox and Rowan 1980). We intervene in the normal processes of the biosphere, interrupting the feedback loops of the process.

Waddington's thesis certainly encompasses science—science being a product of cultural evolution. The function of ethicizing—that is, the development and refinement of ethical systems—is to "mediate the progress of human evolution," which includes the scientific enterprise. No one would argue that the ability to develop ethical systems has been and continues to be conducive to the survival of the species. Obviously, the salient question—for us as citizens and science educators—is, "What kind of ethical systems do we want to evolve?" Do we want an ethic that permits scientists to do whatever can be done simply because it can be done, or do we want an ethic that says, you can do only *a, b,* and *c,* because *x, y,* and *z* are prohibited by universally acknowledged ethical imperatives?

As always, the answer most probably lies in between—and the law will evolve in response to changing societal attitudes. It is a serious mistake, however, to assume that scientists and science, however secular one considers them to be, have no place in the discussions—or have nothing to say about ethics. Science is, and has always been, value-laden. Bentley Glass has stated:

> Somehow there has crept into our writings about the nature and methods of science a dictum that science is objective, while the humanistic studies are subjective, *that science stands outside the nature of man*. What a profound mistake! Science is ultimately as subjective as all other human knowledge since it resides in the mind and the senses of the unique, individual person. It is constrained by the present evolutionary state of man, by the limitation of his senses, and the even more significant limitations of his powers of reason. (Glass 1965)

Public Perception of Science

Part of the problem that faces us as educators is that a segment of the public is still convinced that science is value-free or is supposed to be. The public often feels that science is static, that once the scientific experts have spoken, the issue is settled. Many have been surprised, dismayed, and frustrated when predictions did not pan out or when data required modification.

It is not hard to figure out how public belief in the infallibility of science evolved. The press actively promoted it, and the scientific community did little to discourage it. Many people, unaware of how science actually works and anxious to believe that science and technology would provide the answers to all of society's ills, accepted the notion that experts are infallible, and that scientific pronouncements are value-free. It is interesting that a segment of the public still believes that, thirty-five years after Hiroshima, after two decades of revelations about environmental disasters traceable to excesses of technology, after Skylab, after Three Mile Island, after an about-face on saccharin, and even after a heartrending flip-flop on cholesterol.

Fortunately, surveys continue to show a high level of public confidence in science and scientists. Better still, polls are also beginning to show "the growth of a more realistic public understanding of the limits of technology" (Marshall 1979). For example, a Harris poll conducted in May 1979 and reported in *Science* showed that "in the last three years, Americans have lost confidence in the ability of science to conquer disease" (Marshall 1979). It is amazing how long the notion of science's ability to conquer disease and death has lingered. Here, we have the vast propaganda machinery of the health care system to blame—including the research segment—along with the press and a naive public that does not understand fully the nature of modern medicine, or the nature of the diseases that strike us down. Certainly, disaffection with biomedicine was predictable. The patterns of mortality and morbidity that prevailed in the heyday of the miracle drug and magic bullet are behind us. We in the Western world face a new class of chronic, multifactorial disorders that will not yield to definitive technologies such as antibiotics, immunization, and sanitation. Cancer, heart disease, stroke, diabetes, suicide—all these leading causes of disability and death—require new approaches to prevention and treatment, approaches that address not only the role of the health care system, but also the role of human biology, environment, and lifestyle (Lalonde 1975; Nightingale, *et al.,* 1978; U.S. DHEW 1979; McInerney, Hickman, and Kennedy 1978; American Academy of Arts and Sciences 1977).

Before a disorder will yield completely, however, there must be a thorough understanding of the disease process (Robbins 1979), and although the expectations of the public remain high, that has not been forthcoming for any of the major causes of death in our society. The result is halfway technologies that are mostly palliative, rather than curative or preventive. We have, therefore, diagnostic and treatment regimens that are highly controversial: mammography, electronic fetal monitoring, bypass surgery, and transplantation, for example. And we have, as a result, a debate over the role of technology in the high cost of health care and equally heated debates about the ethical implications of a wide range of new biomedical techniques.

Clearly, science is not an isolated, value-free enterprise even in the laboratory. Scientists undertake certain types of research because their individual values and biases tell them that it is valuable and because the funding of basic research is essentially a political process. Science, however, becomes even less objective when its outcomes are applied to social and public policy. Phillip Handler, president of The National Academy of Sciences, stated in a recent editorial that

> decisions must necessarily still turn on a value judgment, conditioned by social, economic, philosophic, and religious views. But that is the nature of the political process. The public acceptability of a given level of risk—for example—is a political, not a scientific, question. (Handler 1980)

The increased participation of science and scientists in public policy questions has doubtless contributed to the sense of pervasive frustration that finds its most ardent expression in the activities of special interest groups, whether the *cause celebre* is right-to-life, creationism, nuclear power, industrial waste, or recombinant DNA research. The public hears equally well credentialled experts in the same discipline speaking out publicly, usually on the evening news, on different sides of every controversial issue. It is not surprising that people begin to ask, "Who has the valid information? Whom can we trust?"

I quote from Handler again:

> Difficulty arises in the scientific community from confusion of the role of scientist qua scientist with that of scientist as citizen; there is confusion of the ethical code of the scientist with the obligation of the citizen, blurring the distinction between intrinsically scientific and intrinsically political questions. When scientists fail to recognize these boundaries, their own ideological beliefs—usually unspoken—easily becloud seemingly scientific debate.

The confusion that Handler describes, when transmitted to the layperson, contributes to the nagging feeling among the public that the human ability to mediate the environment intelligently has given way to simple manipulation, guided largely by economic and political expedience and by a desire to fulfill individual, personal needs at the expense of long-standing moral principles. So it is not surprising that one hears and reads an increasing number of references to the growth of an anti-scientific, anti-intellectual disposition among the public at large. That is a difficult and distressing problem, given that the direction of research will provide us with more, not fewer, highly controversial issues, particularly as we continue to intervene more and more in basic biological processes.

The Role of the Science Educator

As science educators we are in a position to influence public response to scientific and technological progress. Is frustration at the failures of technology and befuddlement over the often inexact nature of the scientific process to be answered with superstition and unthinking rejection of anything "technological"? Or are we to hope for a healthy, realistic skepticism that fosters good decision-making and active, responsible participation in the evolution of social and public policy questions that involve science and technology (Benne and Barnbaum 1978)?

Answers to those questions are central to considerations of new directions in biology teaching. We in science education must do a better job of teaching the public that science is a dynamic process, that data are mutable, that experts are fallible, and that science and technology indeed have limitations. In their review of secondary biology education in the United States (Hurd, *et al.,* 1980), the authors note changing priorities in the contemporary goals of biology education. While biological knowledge—content—"remains the dominant goal of biology teaching," the authors predict that "the future goal of biological knowledge will place fundamental concepts in a socially relevant and personally meaningful context for students." Indeed, the authors are adamant that "societal issues must be a primary focus of biological education."

What does that imply for the biology teacher? Clearly, it is a call for teachers to provide students with the skills required to participate as effective citizens capable of expressing their own values in a manner that provides at least as much light as heat. Students need to understand that values can and do influence the process of scientific research and its technological applications. They must also realize that complex scientific problems are compounded in the arenas of social and public policy, and that, increasingly, data and analyses from only one discipline cannot provide the answers to our problems and questions. Many of those problems have both their origins and resolutions at the interface of science and values. To imply that science alone will provide the answers is educationally unsound.

Education in Other Arenas

Similarly, those individuals and organizations to whom the public turns for moral and spiritual guidance have a responsibility to acquaint their various constituents with the necessity of hearing all sides in public debates concerning the application of new knowledge. Organized religion, for example, can and should have an important role to play in the development of a scientifically literate public, since many people will continue to turn to organized religion for advice when they must deal with the personal impact of rapid scientific and technologic change. The variety of views contributed by various faiths is vital to our ethical and spiritual well being, just as genetic diversity fuels the fires of evolution and contributes to the biological fitness of the species. But organized religion, if it is not to recede into the protective shell of immutable dogma, must not be afraid to teach its collective congregations to question—to question their own values, the values of scientists and experts, and the values of the religion itself. Blind adherence to dogma is dangerous, whether that dogma has its roots in religion, politics, economics, or science. It promotes anti-rational, anti-intellectual approaches to problems and fosters public acceptance of leaders and their programs who propose simplistic solutions to complex issues. That is a polarizing influence that leaves the middle ground of such issues largely unexamined.

That approach to decision-making, especially in public policy, is directly at odds with the obligations of citizens and government in a pluralistic society. If we are to thrive and prosper together, we must be able to talk to one another and reach compromises in that ill-defined middle ground. To do otherwise is to ensure that public policy will become the province of narrow-minded, self-proclaimed bastions of moral rectitude. Thomas Jefferson wrote almost two hundred years ago that "there is no safe depository of the ultimate powers of the society but the people themselves, and if we think them not enlightened enough to exercise their control with a wholesome discretion, the remedy is not to take it from them but to inform their discretion" (Bazelon 1979). Educators bear a large part of the responsibility for ensuring that society has the skills to counteract the intellectual and political tyranny Jefferson warned against.

The growth and popularity of high school and college courses concerned with bioethics, science policy, and the interrelationships of science, society, and technology (AAAS 1978), indicate the willingness of students, teachers, and administrators to confront those issues in an educational setting. The development of materials for those courses demonstrates that such issues can be addressed in a straightforward and pedagogically sound manner (Hickman 1977; Williams 1980; Commission on the Teaching of Bioethics 1976; Kieffer 1980). The growth and widespread influence of organizations such as the Hastings Center and the Kennedy Institute for Bioethics have done much to "inform the public discretion" at the interface of science, ethics, and religion, and the participation of many prominent scientists, philosophers, and theologians in the programs of those two organizations demonstrates the benefits of interaction between the sciences and the humanities. The range of issues those organizations confront also demonstrates quite well the need for deliberation between the two cultures, such that there is no domination real or perceived by one over the other.

The need for a cooperative effort in that regard was recently acknowledged by a special Rockefeller Foundation Commission on the Humanities that addressed the role of the humanities in American life. The Commission report (Banner 1981) states that "knowledge of the humanities must be coupled with an understanding of the characteristics of scientific inquiry and technological change. Liberal education must define scientific literacy as no less important a characteristic of the educated person than reading and writing."

Much needs to be done if we are to become a scientifically literate public. The work is only beginning. We must improve on our past performance and we must move quickly. The world is growing more crowded, and problems seem to escalate exponentially with the population. Efforts to improve scientific literacy must be cooperative, emphasizing the central role of science without ever losing sight of the importance of human values. We must be mindful of Waddington as we educate those—scientists and nonscientists alike—who will determine how new knowledge will be applied. The task falls to scientists, educators, parents, the clergy, and the media.

The development of new educational programs and materials will be burdened and bolstered by controversy, because we live in a world besieged by complex and controversial problems. The solutions will not be found in an unadaptive, revealed ethic. New times and new problems demand new solutions, a new synthesis of facts, ideas, and values. None of us is naive enough to believe that some long-standing values will not be compromised or altered in the process. But there are always tradeoffs in a pluralistic society. That is the way it is supposed to be if our concern for the commonweal is to be more than lip service. We must all give and take; for the very act of "ethicizing," if it truly has adaptive value, is the ability to transcend oneself, to see something beyond one's place in the present, to appropriate the future. That is a fundamental characteristic of human nature and the one that ensures our survival.

References

ABELSON, P. H. 1980. Science and engineering education. *Science* 210 (4473):965.

AMERICAN ACADEMY OF ARTS AND SCIENCES. Winter 1977. Doing better and feeling worse: Health in the United States. *Daedalus* 106(1).

AMERICAN ASSOCIATION FOR THE ADVANCEMENT OF SCIENCE. 1978. *Resource directory: Programs and courses in ethics and values in science and technology.* Washington, D.C.

BANNER, J. M., JR. 1981. Affinities between scientists and humanists. *Science* 211(4477):9.

BAZELON, D. L. 1979. Risk and responsibility. *Science* 205(4403):277.

BENNE, K. D., and BARNBAUM, M. 1978. Teaching and learning about science and social policy. Boulder, Colorado: ERIC Clearinghouse for Social Studies, Social Science Education and Social Sciences Education Consortium, Inc.

CAPRON, A. M. 1979. Autonomy, confidentiality, and quality care in genetic counseling. *In* Capron, A.M., *et al.* (eds.) *Genetic counseling, facts, values, and norms.* New York: Alan R. Liss.

COMMISSION ON THE TEACHING OF BIOETHICS. 1976. *The teaching of bioethics.* Hastings-on-Hudson: The Hastings Center.

CORPUS CHRISTI CALLER. 1979. Planned Parenthood eyes future fund raising drive. September 14, 1979.

CULLITON, B. J. 1978. Ethics advisory board confronts conception in the test tube. *Science* 202:189.

EDUCATION U.S.A. 1978. Health education: Schools lagging behind demand for change. 20(23):169.

FOX, M. W., and ROWAN, A. N. 1980. A background to the journal and delineation of its scope and goals. *International Journal for the Study of Animal Problems* 1(1):2.

GLASS, B. 1965. *Science and ethical values.* Chapel Hill: University of North Carolina Press.

HANDLER, P. 1980. Public doubts about science. *Science* 208(4448):1093.

HICKMAN, F. M. 1977. Energy and society: Investigations in decision making. Formative Evaluation Report. *BSCS Newsletter* 67:7.

HOLDEN, C. 1979. Teenage pregnancies out of control. *Science* 204(4393):597.

HOLLINGSWORTH, D. R., and KREUTNER, A. K. 1981. Teenage pregnancy. *New England Journal of Medicine* 304(2):121.

HURD, P. D., BYBEE, R. W., KAHLE, J. B., and YAGER, R. E. 1980. Biology education in secondary schools of the United States. *American Biology Teacher* 42(7):388.

HYER, M. 1980. Catholics tighten sterilization ban. *Washington Post*, July 10, 1980.

KELLY, P. T. 1977. *Dealing with dilemma: A manual for genetic counselors.* New York: Springer-Verlag.

KIEFFER, G. H. 1980. Should bioethics be taught? *American Biology Teacher* 42(2):110.

LALONDE, M. 1975. *A new perspective on the health of Canadians*. Ottawa: Information Canada.

LEVINE, C. 1979. Genetic counseling: The client's viewpoint. *In* Capron, A. M., *et al.* (eds.) *Genetic counseling, facts, values, and norms*. New York: Alan R. Liss.

MACINTYRE, M. N. 1977. Need for supportive therapy for members of a family with a defective child. *In* Lubs, H. A. and de la Cruz, F. (eds.) *Genetic counseling*. New York: Raven Press.

MARSH, F. H., and SELF, D. J. 1980. In vitro fertilization: Moving from theory to therapy. *Hastings Center Report* 10(3).

MARSHALL, E. 1979. Public attitudes to technological progress. *Science* 205(4403):281.

MCCOLLUM, A. T., and SILVERBERG, R. L. 1979. Psychosocial advocacy. *In* Hsia, Y. E., *et al.* (eds.) *Counseling in genetics*. New York: Alan R. Liss.

MCINERNEY, J. D., HICKMAN, F. M., and KENNEDY, M. H. 1978. Human genetics: A context for health education. *Health Education* 9(4):33.

NATIONAL ASSESSMENT OF EDUCATIONAL PROGRESS. 1979. *Attitudes toward science.* Denver: Education Commission of the States.

NATIONAL SCIENCE FOUNDATION. 1981. *Science and engineering education in the 80s and beyond*. Washington, D.C.

NIGHTINGALE, E. O., *et al.* 1978. *Perspectives on health promotion and disease prevention in the United States.* Washington, D.C.: Institute of Medicine, National Academy of Sciences.

ROBBINS, F. C. 1979. Assessing the consequences of biomedical research. *In* Altman, S. H. and Blemdon, R. (eds.) *Medical technology: The culprit behind health care cost?* DHEW Publication (PHS) 79-3216. Washington, D.C.

SCIENCE 80. 1980. Wrongful life. December.

SKOOG, G. 1980. Legal issues involved in evolution versus creationism. *Educational Leadership*. November.

U.S. DEPARTMENT OF HEALTH, EDUCATION AND WELFARE. 1979. *Antenatal diagnosis*. Report of a consensus development conference. U.S. DHEW, Publication (NIH) 79-1973. Washington, D.C.

________. 1979. *Healthy people: The Surgeon General's report on health promotion and disease prevention.* DHEW (PHS) Publication 79-55071. Washington, D.C.

________. 1979. Protection of human subjects. HEW support of human in vitro fertilization and embryo transfer. Report of the Ethics Advisory Board. *Federal Register*. June 18, 1979.

WADDINGTON, C. H. 1960. *The ethical animal*. Chicago: University of Chicago Press.

WILLIAMS, J. 1980. *Quality of life and the future*. Final formative evaluation report. Boulder, Colorado: Biological Sciences Curriculum Study (unpublished).

Commentaries

McInerney's paper is provocative. I feel all science teachers could benefit from reading it. I agree that scientists should be involved in the development of public policy and that citizens should be informed on the nature, capabilities, and limitations of science. It is unfortunate that social, ethical, and legal progress lags well behind that of science and technology. The almost unbelievable inroads creationists are making in some states—as well as the danger of mutagens, carcinogens, and teratogens in our modern environment—should make us all realize the importance of science education in developing good decision-making skills and a willingness to participate actively at the science-values interface.

Jane Abbott

"Science is a method of obtaining new knowledge when the method used and the knowledge thusly gained are objective, verifiable, and reproducible." This dogma is probably how the general public understands science and its predictable uses through technology. That no consideration is ever given to the consequences of scientific endeavors is *a priori* in the minds of most laypersons. Therefore, we see the "anti-scientific, anti-intellectual" backlash we are now experiencing with inadequate funding and misplaced priorities.

We as science teachers must bear part of the blame. Many of us are products of the NSF-funded "honeymoon" of the past decades. This vast support has engendered a self-righteous, elitist arrogance that believes it is devoid of subjective valuing, but is, in fact, tainted by its own subjectivity. We are products of an era that ventured into space and landed a man on the moon only ten years after we decided that we should. We took the predictions of the Watson-Crick Model and developed bacteria capable of synthesizing human insulin. We reduced the size and cost of calculators while increasing their capabilities tenfold in the same short period of time. We believe that research and technology are the cure for all our ills. We erroneously believe that values and valuing do not guide the course of our technical progress, but rather follow it.

Therein lies our problem. We must reorient our focus as science teachers and recognize that worthwhile progress comes through the careful application of the knowledge gained through science to the needs that people identify. We need to define our own personal values in order to recognize when our teaching is value-laden and to measure the degree to which *how* we are teaching affects *what* we are teaching. Only when we recognize our own blind adherence to dogma will we be able to transcend ourselves and help others "to see something beyond one's place in the present."

Roy K. Baer

Our task as science educators can best be seen in light of that oft-used quotation by British clergyman and poet John Donne: "No man is an Iland of it selfe; every man is a peece of the Continent, a part of the maine."

We must recognize that, as McInerney states, if we are to thrive and prosper together, we must be able to talk to one another and reach compromises in that ill-defined middle ground. If we as science educators are in a position to influence public response to scientific and technological progress, we must then bear a large part of the responsibility for developing a scientifically literate society. Again as so beautifully stated by McInerney: liberal education must define scientific literacy as no less important a characteristic of the educated person than reading and writing.

Life then is but a constant handling of tradeoffs. It is the ability of the individual to handle these tradeoffs, in light of established values, that marks a well-educated individual capable of dealing with the multitude of interrelationships needed for survival.

David P. Lapierre

It is not unusual for high school students to reflect long-standing ignorance and prejudice, not only in the realm of scientific knowledge, but in the social and ethical domains as well. Their values, poorly defined and understood, have often been "kept separate" from scientific data and technology. Neither area has developed as needed because the areas have not met in the same light.

Early in their educational development, students need experiences that will help them realize that there is subjectivity in all areas of learning and that intelligent skepticism is an essential ingredient of progress. It is a point from which confident steps can be taken when students are no longer needlessly frustrated by the realities of fallibility, flexibility, and lack of finality. It is necessary that they "save their frustrations" for the processes involved in courageous approaches to complex problems now and in the future.

As indicated in McInerney's paper, the task of educating both professionals and the general public lies with scientists, professional educators, parents, clergy, and the media. Interaction is imperative.

Sister Marian Catherine McGrann

It is true that scientific literacy is necessary both for those who will become science professionals and for those who will not. Both groups will decide what laws and public policies will be formulated. Much of the anti-intellectual and anti-scientific disposition of the public has arisen because of our past inability to communicate adequately the nature, limitations, and capabilities of both basic and applied science. The general public still has considerable confidence in science and technology, but this confidence is weakening—shaken by "failures" of science and technology to solve our seemingly endless array of problems today. McInerney's paper would provide an excellent focus for discussion and seminars in workshops and departmental meetings. Some educators still teach science as a value-free activity. They might benefit greatly from focusing on science *and* society in their lesson planning and curriculum development.

James A. Meyer

Clearly, the consideration of human values in today's science classroom is necessary to promote a better understanding of science and its relationship to the world. Equally as clear, however, is the need to distinguish *science* from values. I might contend that science remains objective, while scientists and laypeople are subjective. What people do with the "facts" garnered through scientific investigation is oftentimes value-oriented. Two people can, and do, see the same fossil evidence and conclude divine intervention or gradual evolutionary processes.

In a state such as Oklahoma, which has a conservative reputation and a strong religious movement, the discussion of value-laden topics arises not from my prompting, but from my students' dogmatic view about the world. The rub arises when what is taught conflicts with what is believed. In many cases the beliefs win out, and student and professor go happily along after the class still promoting their own opinions. I spent a week on the principles of evolution in my introductory *botany* class with all examples taken from the plant kingdom. On a midterm I asked for an example of evolution, and several students gave the old, and incorrect, ape to man sequence! Where did I go wrong? Obviously, a rapid acceptance of new ideas is hardly the normal course of events in religious activities or when dealing with religious doctrines. Value-laden discussions benefit mostly those who have no previously formed opinions.

A bill that would have forced all public school teachers in Oklahoma to give equal time to the creationist view was recently defeated in committee, not because separation of church and state was invoked nor because the case for evolution was well presented. Instead, it was argued that in most Oklahoma public schools in which evolution is taught today, creationism is the *only* theory that is proposed to the students. Thus if the bill passed, Darwinian philosophy would be given more time in the classroom than before. Also, some legislators voiced protests that this would, in their opinion, reduce the status of creationism to a theory from that of a fact. This is hardly a compromise, nor an intelligent exchange of information. Human values will enter the classroom, but must *I* be objective about them?

Gordon E. Uno

8

THE CONTROVERSY ABOUT CONTROVERSY: ISSUES OF SCIENCE AND SOCIETY IN THE CURRICULUM

Faith M. Hickman

> *Education in the social and life sciences . . . is facing new territory as yet uncharted. That territory is circumscribed on one boundary by the rapid explosion of human knowledge and on the other by the inability of human systems and societies to respond rapidly in the realms of values, ethics, and public policy.*

Many educators believe that—in order to achieve the kind of active, reflective, politically sophisticated citizens that a democracy requires—classroom instruction cannot, indeed must not, avoid full exploration of potentially controversial issues. Others—including some parents, teachers, and school administrators—disagree This situation can be described as "the controversy about controversy." The issue of whether to teach about controversial issues is, in itself, controversial. Issues of science and society are often controversial. The decision to deal with them (or not to deal with them) in the classroom must be carefully made.

This paper explores the "controversy about controversy" through an examination of (a) the definition of controversial issues; (b) the arguments and counterarguments used to favor or oppose such issues as legitimate subjects within the school curriculum; and (c) the implications for the classroom teacher of the contemporary position that favors the inclusion of controversial topics in the student's school experience.

IN 1978, a consortium of twelve professional education societies, ranging from the Music Educators National Conference to the National Science Teachers Association, distributed a policy statement on what they perceived as the essentials of education. Their statement began:

> Public concern about basic knowledge and the basic skills in education is valid. Society should continually seek out, define, and then provide for every person those elements that are essential to a productive and meaningful life. . . .
>
> . . . Educators agree that the overarching goal of education is to develop informed, thinking citizens capable of participating in both domestic and world affairs. . . .
>
> . . . the essentials of education include the ability . . . *to make informed value judgments*. (emphasis added) (Organizations for the Essentials of Education 1978)

Today there is little argument among educators (at least in public) that education for informed citizenship and enlightened decision-making is among the primary goals of schooling in the United States. However, there is often dissension among educators and between educators and the public (including parents) that such a goal can or should be sought. The problem is not so much the end—as the *means* to the end. Many educators believe that—in order to achieve the kind of active, reflective, politically sophisticated citizens that a democracy requires—classroom instruction cannot, indeed must not, avoid full exploration of potentially controversial issues (BSCS 1978a, p. 21). Others—including some parents, teachers, and school administrators—disagree. This situation can be described as "the controversy about controversy."

Thus, the issue of whether to teach about controversial issues is, in itself, controversial. The controversy can and does occur in any area of the curriculum, but issues of "science and society" frequently make headlines. For example, debates continue over whether to teach about such emotionally laden topics of science and society as abortion, birth control, endangered species, environmental protection, genetic engineering, evolution, human sexuality, and a host of others. And even when parents, school administrators, teachers, and students can reach some consensus on the value of instruction about controversial subjects, debates continue about how best to handle such difficult topics. Horror stories and lawsuits are sometimes the results of well-meaning but misguided educational experiences (Freeman 1978).

In order to better understand this "controversy about controversy," we need to examine three areas: (a) the definition of controversial issues; (b) the arguments and counterarguments used to favor or oppose such issues as legitimate subjects within the school curriculum; and (c) the implications for the classroom teacher of the contemporary position favoring the inclusion of controversial materials in the student's school experience.

What is a Controversial Issue?

Controversial issues are easier to identify than they are to teach about. In 1963, the National Education Association committed itself to the teaching of controversial issues with a formal resolution based on the following definition:

> A controversial issue involves a problem about which different individuals and groups urge conflicting courses of action. It is an issue for which society has not found a solution that can be universally or almost universally accepted. It is an issue of sufficient significance that each of the proposed ways of dealing with it is objectionable to some sector of the citizenry and arouses protest. The protest may result from a feeling that a cherished belief, an economic interest, or a basic principle is threatened. It may come because the welfare of organizations or groups seems at stake. When a course of action is formulated that virtually all sectors of society accept, the issue is no longer controversial. (p. 153)

A similar position was taken fifteen years later by a curriculum development organization interested in issues of science and society.

> Issues become controversial in human society when no consensus can be reached on the solution to a problem. Differing theoretical perspectives, beliefs, and other patterns of thought among people result in differing points of view. Such differences grow out of the strong interests, the deep commitments, and the close group affiliations represented within a community. The bitterness generated by a particular controversy depends on the intensity of the emotions aroused. . . . Predicting what will or will not arouse emotions in our society is virtually impossible. (BSCS 1978b, p. 33)

There are, of course, many issues that can be classed as controversial or potentially so. No element of the curriculum is immune. English and American literature can arouse dissension, as evidenced by the Kanawha County, West Virginia textbook controversies. Art, music, history, and even such precise and revered subjects as chemistry, physics, and mathematics can generate controversies in different times and different places. It is not the subject matter, but the interpretation that is given the subject matter, that causes the controversy. In this discussion, issues of science and society that most directly confront teachers of life science and social studies will be emphasized. But the themes are the same, no matter what element of the curriculum is considered. As a point of reference, however, it is useful to consider briefly some of the issues that life science and social studies educators and authors have mentioned recently in writing about controversial issues in the classroom:

> The issues . . . are numerous and diverse: nuclear power, abortion, the Equal Rights Amendment, hunting of whales and seals, *in vitro* fertilization, the snail darter *versus* the TVA, sex education, euthanasia, and the future of the Alaskan tundra. (Hickman 1979)
>
> . . . some maintain that teaching about reproduction may weaken morals, promote promiscuity, and result in unwanted babies or V.D.; or teaching about the environment may challenge the practice of those who pollute or those who unwisely exploit our natural resources; or teaching about evolution may undermine religious beliefs or what some perceive as our cosmic importance. (Moore 1979)
>
> Many . . . interactions are controversial. The critical issues of energy resource allocation, the use of nonrenewable resources, population growth policies, support for scientific research, genetic manipulation, human rights in scientific investigation—indeed, human rights themselves—involve social and ethical dimensions. (BSCS 1978b)

The 45th Yearbook of the National Council for the Social Studies (Muessig 1975) was devoted to controversial issues. Chapters to illustrate specific areas of controversy included sex modes and values, death and dying, majority rule, integration as a societal goal in a pluralistic nation, "the planet in peril," and the nation-state versus world organization.

It is relatively simple to catalog issues that are controversial now although it is difficult to predict what such issues will be in the future. But among the most controversial of all is the question of whether to teach about such issues in the nation's classrooms. In 1848, Horace Mann summarized the view of his time: that schools should teach the common elements that all can agree to, but should avoid controversial topics (Butts 1978, p. 94). Over a century later, the NEA (1963) took a radically different stand:

> Rational discussion of controversial issues should be an important part of the school program. The teacher should help students identify relevant information, learn the techniques of critical analysis, make independent judgments, and be prepared to present and support them. The teacher should also help students become sensitive to the continuing need for objective re-examination of issues in light of new information and changing conditions in society. (p. 175)

Most contemporary professional literature in education supports the NEA view. However, groups of political conservatives or religious fundamentalists ardently refute such a stand. Notable in this regard are the Gabler and McAuley families in Texas, the Creation Research Institute, the Heritage Foundation, and the Moral Majority. Teachers may feel trapped between opposing points of view and may be tempted simply to avoid potentially controversial material in the classroom. "The teaching of facts, as a means of avoiding controversy, is safe" (Hoover 1976, p. 148).

Safety has its merits and its rewards, but it is only one factor in the complicated equation we call education. Before choosing the safe option over all others, educators need to make rational assessments of the demands that controversial issues place on teachers, students, and schools. Such assessments can begin with a review of the nature of the "controversy about controversy" and of the various arguments that have been used to favor or oppose the inclusion of such topics in the curriculum.

Perceptions and Definitions of Curriculum

Eggleston (1977) asserts that "(u)nquestionably the political debate about curriculum is now a permanent, institutionalized feature of Western educational systems" (p. 142). He views this situation as growing, at least in part, from two well-established and mutually exclusive models of what curriculum is or should be The *received perspective* "is one in which curriculum knowledge, like other components of the knowledge system in the social order, is accepted as a received body of understanding that is 'given,' even ascribed, and is predominantly non-negotiable. . . ." The *reflexive perspective* ". . . is one in which curriculum knowledge, like the components of the knowledge system, is seen to be negotiable; in which content may be legitimately criticized and argued or new curricula devised. . . . (I)t is . . . manifestly subject to political and other influence; a construction of those who participate in its determination" (p. 52).

That the controversy about controversy grows out of dissimilar views of knowledge is recognized at the most fundamental level: ". . . (one) kind of issue involves science as a way of knowing. Issues of that sort arise when science offers explanations of such things as human origins, organic evolution, cosmology, disease, life, and death that are incompatible with those arrived at intuitively outside of science, or through theological or philosophical systems of thought" (BSCS 1978b).

Perhaps no specific case better illustrates this fundmental debate than the evolution-special creation controversy which continues to rage fifty-five years after the Scopes Trial in Tennessee:

> But scientific matters are not resolved by democratic procedures. Democracy did not give us the laws of gravitation, the laws of thermodynamics, or Mendel's laws of inheritance. In a science class, creationism is not—indeed cannot be—a part of science because its statements are, in the last analysis, based on revelation, not on a careful marshalling of data by observation and experimentation. Creationism is religious dogma; evolution is scientific theory. Thus, scientists should oppose the teaching of creationism as science, though no one should object to it or any creation myth being taught as part of the history of religion. (Moore 1979)

This passage and others like it are good proof of the old maxim that politics makes for strange bedfellows. The view espoused by Moore and promoted by all major professional societies in biology and biology education is that creation has no place in the science curriculum. On *this* issue, scientists and educators seem to be approaching curriculum from the *received* perspective—in that course content is a given, in this case, because of the nature of science as a discipline. Curriculum is, in this case, non-negotiable, because science is science and religion is not. The counterarguments such as "equal time" or "scientific creationism" appear rooted in the *reflexive* perspective: that curriculum is a human construct open to political influence and determination.

The paradox arises when the views of the two groups are compared on matters other than evolution. For example, the National Association of Biology Teachers, while opposing the inclusion of Genesis in the science classroom, nevertheless favors inclusion of issues of science and society in the curriculum (NABT 1979). This is an essentially reflexive view. On the other hand, supporters of the creationist point of view often argue from the received perspective when they support emphasis on "basics" and other "traditional" subjects in the curriculum.

The NABT has recognized and attempted to resolve the seeming discrepancy between its stand on the evolution issue and its commitment to issue-oriented education by adopting a policy statement:

> The credibility and usability of science depends upon maintenance of the integrity of science as a discipline. While no feature in this policy is to be construed as preventing the full range of applications of science and the elucidation of its social and humanistic implications, there is an obligation to insure that the scientific data thus used is both accurate and derived within the accepted procedures of the discipline. (NABT 1980)

The point of this analysis is that, whatever issue is confronted, its import lies not in the issue itself, but in the ideological and political perspectives that are taken. Curricula may be viewed as received, reflexive, or both, depending upon the situations, the persons involved, and the social and political climates in which individuals and groups operate.

Course Content and the Distribution of Power

Controversies about curriculum arise from ideological differences. They also arise from concerns about the distribution of power among all the individuals and groups who perceive themselves as having some vested interest in the schools.

Eggleston (1977) describes a situation in which a young teacher undertook an extensive program of community-based education. Students who had previously exhibited antisocial behavior established and ran a playground, dug gardens for senior citizens, painted, cleared away litter, and brought noticeable improvements to a rundown area. The students and the teacher were praised by parents and the local press. But activities did not stop there. The class found that the old people had difficulty obtaining rent rebates, that the local council was late in fulfilling a promise of roof repair, and that blocked drains and other problems with municipal services were beyond the power of the students to solve. The class decided that action was needed. They wrote to the housing committee and to the newspaper urging immediate action. The result? They were condemned by both the local government and the press as "teenage troublemakers led by a radical teacher." Thus, Eggleston concludes: "Here we have reached one of the fundamental truths of the reflexive perspective: that the curriculum is about the distribution of power" (p. 83).

Thus, controversies arise about the inclusion of controversial subject matter in the curriculum (or about the kinds of activities that students should proposedly be engaged in) because some individual or group perceives a threat to some authority already held or some unfair or frightening redistribution of power. This power struggle may occur between parents and school personnel, between school personnel at different levels (for example, school boards and teachers), or between schools and some form of government, either local, state, or federal. Such disagreements also give power to a whole array of "screeners" or "gate keepers"—either officially or self-appointed "protectors" of curriculum content.

Parents, although repeatedly charged with apathy and neglect, continue to express a genuine interest in the schooling of their children. As individuals or collectively, parents claim the right to have some say about what goes on in schools. They demand some power for two reasons: they produce the clientele and they pay for the system (Roper 1977). Parents may argue that, because they are paying for the schools and because it is their offspring that are being educated, they should have the power to veto inclusions in the curriculum that they find objectionable for any reason. Examples of this kind of power struggle are easily found, in areas ranging from sex education to the evolution-creation controversy.

Disagreements may also arise among teachers and other school personnel. This kind of struggle is rooted in the ideological perspectives discussed previously, but is usually manifested in some power struggle. A good example is the recent court case of *Dale v. Board of Education* in South Dakota. In this instance, a school board had insisted that Mr.

Dale spend less time treating the topic of creation and give more time to what the board perceived as other important topics in biology. The court chose to rule, not on the issue of teaching about creation, but rather on the question of whether due process procedures had been adhered to and the teacher treated fairly. The court found that the appellant's rights of due process had not been violated (Brandenburg 1980). Thus, the court upheld the power of the school board to make demands on curriculum content, so long as due process was given.

The struggle for power is also evident in the interactions of schools and government, either local, state, or federal. For example, one school administrator predicts trends in education over the next twenty-five years and comments on the implications of his predicted trends:

> *Fact:* The educational system will receive a large portion of funding from the federal government and will more frequently become a mechanism for carrying out government policy.
> *Likely Result:* There will be more required aspects of the curriculum. Concern will be voiced about academic freedom and government control.
> *Possible Implications for Education:*
> 1. Courses to implement national priorities will be instituted. Areas covered will be energy conservation, birth control, genetic counseling, an anti-alienation and antiviolence curriculum, and personal and national security.
> 2. The national teacher organization will resist further government-mandated programs. (There will be only one teacher union.) It will be joined in this fight by a coalition of school boards, school administrators, the national Chamber of Commerce, and labor unions. (Miller 1978)

Miller's predictions may not prove true, but we can be assured that power struggles will occur. If past trends continue, such struggles are more likely to be resolved through self-censorship than through the courts. Some authors see our current self-screening practices as a potentially disastrous threat to free education:

> A . . . serious problem arises when teachers are prevented from presenting some idea, not because it is wrong, but because it will offend, challenge, or demolish a firmly held belief or way of life. . . . Restraints . . . occur frequently enough to keep schools alert to the possibility of protests from irate parents, and they have a truly chilling

effect on textbook publishers (Fitzgerald 1979). Publishers are in business to make money, and they do not make money by not selling their books. Rarely, it seems, do they ask whether they might sell more books with a balanced and judicious presentation of controversial topics than by avoiding such issues entirely. Thus, textbook publishers contribute to a serious and pervasive censorship. Their books tend to be bland, out-of-date, and thoroughly safe representations of topics to which no one can possibly object.

Once screened by publishers, the books are examined by school boards. Responding to the same pressures, these groups usually select the safe books and often specify the curriculum.

The third group of screeners is classroom teachers, though at this point, not much variety is left to be screened. This group is the one that faces the real problems: the drug scene, decline in test scores, violence, integration and busing-related problems, accountability, reluctant learners, declining enrollment, mainstreaming, and society's reluctance to assign students to groups according to ability. Considering all these problems, one can understand and sympathize with the main concern of many teachers today—survival. (Moore 1979)

Risk versus Benefit and Values versus Indoctrination

Some of the most often heard arguments against the treatment of controversial issues in the schools fall within two broad categories that may be defined as risk versus benefit and values versus indoctrination. The risk/benefit analysts argue that too much is to be lost: that school time is limited and that valuable material must be excluded if controversial topics are introduced. Others argue that school students haven't the maturity to deal with problems so complex that even their elders cannot find resolutions. So the benefit, in this case, is seen as minimal or perhaps negative (if students become confused or fearful), while the risk taken in turning aside the proven value of traditional education is unacceptably high (NEA 1963, p. 154).

Proponents of the study of controversial material take a different view:

> We live in a controversial and litigious age, and our students constitute the citizenry who will decide the fate of many issues based on biological data. If we treat the classroom as an isolate from societal problems we fail to prepare students to clarify their values and make judgments on the basis of the best information available. (Mayer 1980)

Also,

> But how are young people supposed to learn to "handle themselves" without participation in situations that require them to do just that? How are they to develop self-confidence in decision making without ever facing a decision? Can abilities for action at some time in the future arise from inaction in the present? (Hickman 1979)

Another broad class of controversies about controversy rests in the distinction between "values education" and indoctrination. Here, both camps find some common ground, although the boundaries of that ground may be surveyed again and again with differing results. In discussing the limitations and problems of "value-focusing activities," Hoover (1976) stressed the following points:

> Basic substantive values (e.g., one's feeling toward religion, politics, and the family) are usually respected as private, personal affairs. One may raise questions, designed to provoke reflective thought, but he must not provide the "right" answers to such questions.
>
> Teachers naturally hold values that are highly cherished. By word and deed they tend to reflect such values. Such activities can hamper effective instruction in the realm of values. (p. 148)

A classroom teacher provides a concrete example of concern over the inculcation, rather than the exploration, of values:

> I am very careful not to interject my personal bias. I think it important that students are given ammunition to make their own intelligent decisions. For example, during the Dr. Steptoe "test tube" baby controversy, the social implications of the "biological advance" were discussed in my classroom. When a student said, "That's disgusting," I would simply ask, "Why did you say that?" This would elicit further discussion aimed at developing the pros and cons of the role of scientific knowledge in society. (Resnick 1981)

Students' Rights and Education for Democratic Participation

To be sure, debates over curriculum philosophies, the distribution of power, risk/benefit, and indoctrination have been prominent in the controversy about controversy. But the contemporary view that holds sway in nearly all the professional literature, and the arguments that are consistently used as the greatest weight of evidence supporting the inclusion of controversial subject matter in the curriculum, are based on two major ideas: students' rights and the function of education in a democratic society. In 1963, the NEA reported:

> In recent years many groups of professional educators, academic scholars, local school boards, and lay citizens have adopted the view that the study of controversial materials and issues is an inherent right of pupils and a way for them to learn about significant aspects of their culture. This view has been incorporated in policy statements of many boards of education in terms such as this:
>
> The pupils of Armijo Joint Union High School District have four rights to be recognized (in conjunction with study of controversial issues):
>
> (1.) The right to study any controversial issue which had political, economic, or social significance and concern which (at his level) he should begin to have an opinion. . . .
> (2.) The right to have access to all relevant information, including the materials that circulate freely in the community.
> (3.) The right to study under competent instruction in an atmosphere free from bias and prejudice.
> (4.) The right to form and express his own opinions on controversial issues without thereby jeopardizing his relations with his teacher or the school.
>
> (California State Department of Education 1960)

Nearly two decades later, another group working independently and without knowledge of this previous statement, developed an almost identical list:

> Learners have certain rights that must be considered in establishing policies for the study of controversial issues in the public schools:
>
> 1. They have the right to study those controversial issues that are appropriate to their level of maturity. In general, those issues that learners of a particular age are beginning to form opinions about are appropriate to that maturity level.
> 2. They have a right to obtain relevant information freely.
> 3. They have the right to seek information in an atmosphere as free from bias, prejudice, and coercion as possible; bias where it appears should be clearly labeled.

4. They have a right to form and express opinions on controversial issues without jeopardizing their relations with teachers or peers. (BSCS 1978b)

Such startling similarities can only be explained as a function of virtually identical premises about a democracy. Yet the pervasiveness of such premises across time and distance should not be surprising. Actually, the argument of education for active participation in a democracy pre-dates both groups. The roots of American education as a mechanism for promotion of liberty, equality, and the public good can be traced back to the American Revolution:

> The abstract goals of the projected political community described in the Declaration of Independence, the Preamble to the Constitution, and the Bill of Rights could easily be regarded as the political values that public education ought to promote. . . . Thus the public and educators alike were left with the problem of designing systems of education that would be able to promote the common values of an as yet inchoate national political community. . . . (Butts 1978, pp. 22-23)

But the most eloquent portrayals of education for democracy stole center stage after World War II and continued in a starring role into the 1960s. Alexander Meiklejohn (1960), teacher, philosopher, and spokesman for civil liberties, wrote:

> . . . the protection of public discussion in our nation takes on an ever-increasing importance as the nation succeeds in so educating and informing its people that, in mind and will, they are able to think and act as self-governing citizens.

The Supreme Court acted on just such a view in a 1952 ruling (*Wieman v. Updegraff*) asserting that teachers have a special obligation to promote freedom of thought:

> That our democracy ultimately rests on public opinion is a platitude of speech but not a commonplace of action. Public opinion is the ultimate reliance of our society only if it be disciplined and responsible. It can be disciplined and responsible only if habits of open-mindedness and of critical inquiry are acquired in the formative years of our citizens. . . .

> To regard teachers—in our entire educational system, from the primary grades to the university—as priests of our democracy is therefore not to indulge in hyperbole. It is the special task of teachers to foster those habits of open-mindedness and critical inquiry which alone make for responsible citizens, who, in turn, make possible an enlightened and effective public opinion.

Full explication of this idea is impossible here, because the volumes of writing are too large to summarize or even credit. On this particular issue, many contemporary educators view the case as closed. Education for participation in the processes and problems of a democracy is a *sine qua non* for the schools:

> In a free society, citizens who do not know what they believe are a danger to the society and therefore to themselves. . . . In order that values may be clear, the methodology of value formation and reformation should be taught. Respect for cause-and-effect relationships, testing of one's present values against evidence, and the origin and development of a philosophy of life are but three of the matters to be treated in a school where creation and re-creation of value structures are a concern. (Doll 1974, p. 90)

Another author summarized the argument:

> I should point out that education is never an objective, unbiased, value-free endeavor. We fool ourselves if we assert that education is pristine, noble, or untarnished by the vapors of the political arena. It just is not so. The function of education since the beginning of time has been to create a certain kind of person. The kind of person a particular society desires to create will change over time and will differ between and among societies. But the basic goal of "molding young minds" does not vary. . . .
>
> Educators are now making explicit commitments to education for informed and enlightened participation in the resolution of issues of public policy. That means that educators are coming to believe that students should leave their school experience capable of understanding the biosocial issues of today and tomorrow, informing themselves about such issues as they emerge, and evaluating the many points of view that are heard on such issues. Students should be capable, if they choose to do so, of participating in the political processes that lead to the development of public policy on such issues. (Kennedy 1980)

Implications for the Classroom

No matter how compelling the arguments for students' rights and education for democratic participation, the decision in the end falls to the classroom teacher: "Even though the legislature may decree (at the societal level) that the values of society should be taught in the schools, . . . the skills, knowledge, and attitudes of the teacher are a key in the final determination of what is taught and how" (Goodlad and associates 1979, p. 185). Practical constraints are myriad; support mechanisms are few. Some, like Mary Breasted (1970) when she writes about sex education, take a dim view of what the future may bring:

> But I think it will take a long time before the high schools start telling their charges the whole truth. . . . So the young will go on being cheated out of the truth. But then, they're used to that. (p. 336)

Others are more optimistic:

> It is unquestionably difficult . . . to give students an understanding of the basic scientific concepts in a field and also to provide enough relevant information to enable them as future citizens to deal intelligently with difficult political, economic and social issues. But progress can be made; students can begin to develop critical standards that will help them to sort out and appraise the technological claims and advice they receive through the popular media. This task has not been given sufficient attention in past curriculum development efforts and needs to be readdressed. (Panel on School Science 1979, p. 9)

The pressures that may operate upon any teacher who is trying to decide about the inclusion of potentially controversial subjects in the curriculum are derived from the arguments and counterarguments summarized earlier. Albury (1980) has identified two key difficulties that prevent easy integration of such studies into the school day. The first is that every issue has a political dimension. "No matter how non-political the science teacher attempts to be in discussing such issues, he or she is always open to the charge of 'playing politics' in the classroom. As Jean-Jacques Rousseau once said, 'You may not be interested in politics, but politics is interested in you!'" The second serious difficulty, according to Albury, is lack of technical expertise. The teacher cannot be expected to have mastered all the complexities of the scientific knowledge that may underlie any particular controversial issue. Nor can the teacher be expert in the principles of morality, ethics, and philosophy that are so important to the dissection of such issues. But this problem is a double-edged sword: ". . . the only people who seem to be qualified to discuss the issues in question are technical experts of various sorts. But unfortunately, most of the ethical controversies relating to science and technology involve controversial issues for the experts as well. . . ."

Although the teacher may be no better qualified to actually make a decision than any other citizen, the teacher *should* be qualified to assist students in the analysis of the issues and of the nature and extent of the available options. Thus, teachers and students should work together for better understanding, rational discourse, and reflective decision-making.

It is true that the decision of whether to deal with controversial topics is one that every teacher must make. Too often, the decision is made by default. Teachers skirt the issues, plead lack of time, or find solace in the facts of the discipline to the exclusion of more demanding questions. The decision *not* to deal with controversial issues can be legitimate, so long as it is made carefully. Teachers should assess candidly the interests and concerns of their students. They should explore the attitudes and values of school administrators and fellow teachers. They should study the desires of the community and seek out avenues of cooperation and assistance between and among school personnel, parents, and the general public. The decision to teach or not to teach about controversial topics should be as carefully and thoughtfully made as any other decision.

Furthermore, it is a decision constantly open to reconsideration. It cannot be made once and left to rest for one's entire teaching career. Rather, each new school year and each new learning opportunity should stimulate re-examination of one's commitment either to the treatment or the avoidance of a single issue or issues in general.

Practical Considerations

Once the decision is made to deal with a controversial issue, there are certain procedures and practices that have proven valuable in making the instruction acceptable, meaningful, and worthwhile. A full discussion of all the "how-to's" is beyond the scope of this paper. The extensive literature of values education is recommended (see the Selected Bibliography on Values Education). Also useful are the *Biology Teacher's Handbook* (Mayer 1978) and *Controversial Issues in the Social Studies* (Muessing 1975). There are, however, certain key elements of instructional practice that are so basic and so pervasive that they demand mention here.

Community Involvement

Most all those who favor the inclusion of controversial material in the curriculum and who offer advice to teachers stress the importance of informing and involving the community. Freeman's advice (1978) on education about death and dying could apply equally well to other issues:

> A teacher planning a unit on death and dying should realize that he is dealing with controversial material and take certain precautions to protect himself and his students. He should discuss the proposed class with the principal or curriculum consultant who may have a feel for possible community reactions. He might also plan a parents' meeting to explain the goals and activities of the class and to solicit parental input as to course content. Finally, for the protection of all, the teacher should require a permission slip from each participant's parents. Parents will know better than the teacher whether or not their child has the emotional maturity to handle the subject matter.

Gauld (1980) concurs:

> Prior consultation with parents about topics and activities likely to be considered controversial may help alleviate some of the concern parents sometimes feel when schools appear to be stepping beyond the bounds by which one might legitimately expect them to be constrained.

The NEA (1963) takes this concern one step further, by advocating school board policy statements that protect teachers and school administrators from unjust pressures from partisan groups.

> One means of providing this protection is for the board of education to develop a specific written policy to govern the teaching of con-

> troversial issues. The teacher who works responsibly within such a framework has assurance of support . . . should unfair attacks be made. (p. 161)

Selecting Issues for Consideration

Given limited classroom time and a variety of instructional objectives, it is impossible to provide substantive instruction on all the issues that might be considered relevant to students in science or social studies classrooms. How is the selection of a particular issue for study to be made? Students' interests and concerns play an important role. So also do the background and interests of the teacher. The NEA (1963) provides a list of criteria for consideration:

> Is the issue suitable for pupils of the maturity and backgrounds represented in the class?
>
> Will study of the issue help students achieve course objectives?
>
> Is the issue one for which adequate study materials that present various points of view can be obtained?
>
> Is the issue one which is important and likely to be of continuing significance?
>
> Is the issue one that the teacher is—or can be—adequately prepared to handle fully and objectively? (pp. 156-158)

Again, we find some uncanny similarities between this list and that developed by the BSCS (1978b) fifteen years later. The BSCS statement echoes these criteria for decision-making, while adding two more:

> Can means be provided for the evaluation of student outcomes related to the study of controversial issues?
>
> Can the issue be appropriately considered within reasonable classroom time?

Teaching Styles and Strategies

For most teachers, the questions of community involvement and curricular decision-making are of less immediate concern than problems of teaching styles and strategies. The prospect of approaching thirty students with controversial material can be frightening. The teacher may ask: How can I be sure to achieve a balanced presentation?

How can I keep students, whose emotions may become feverish, on a productive train of reasoning and discourse? How should I respond to questions about my own values?

Baker and Doran (1975) describe eleven alternative teaching techniques.

> These range from those which identify value issues in some of the topics dealt with in science and related to the pupil's awareness of these issues, through those which ask a student to tease out the knowledge and value components in the proposed solution to a problem and to consider and compare alternative solutions, to those which require a student to express a choice or to make a decision following the analysis. (Gauld 1980)

Whatever technique is used, certain steps can help achieve full participation and reflective discussion. Kennedy (1978) recommends a seating arrangement that will promote interaction. Students can be organized into groups of three to five and asked to describe different points of view about the information that has been presented. Groups later share their findings with the rest of the class. He cautions that all students should have an opportunity to express their ideas and feelings.

He also advises the teacher to remain neutral. The teacher may be tempted, or feel compelled, to point out the "right way of thinking" but such an approach cannot promote the student's own intellectual growth. Teachers should use unbiased, careful questions to encourage students to examine critically all views that are presented. Silences, however long, should be accepted as a natural part of the process of reflective thought. Since divergent points of view are to be respected and valued, the teacher should avoid comments indicating either approval or disapproval. Nevertheless, the student's willingness to contribute should be honored with comments such as "Thanks for that idea," or "I never thought of it that way."

Kennedy also suggests that the teacher should display some measure of enthusiasm and personal familiarity with the issue. The teacher should respect students' individual differences and should receive all responses graciously and without derision. The teacher should avoid: giving information that students can obtain from written material or from each other; permitting a few students to dominate the discussion;

rejecting a seriously offered idea; responding negatively to ideas with which he or she may disagree; and standing in front of the class as an authority figure.

Emotionalism/Rationality

The goal of education about controversial issues is intellectual, not emotional. It is not the business of educators to promote either firebrand rhetoric or shared ignorance. Therefore, the teacher must be selective in the learning materials that are used and well prepared for the ensuing discussions. Again, the NEA (1963) has provided a useful set of criteria for the selection of materials:

> 1. Are the materials adaptable to the range of reading ability among students who will use them? Some material should be easier and some more advanced than the average reading level of the class to provide for the inevitable range of individual differences.
> 2. Is there in the collection a balance among various points of view on the issue?
> 3. Are some objective accounts that are balanced within themselves included? If not, is it because none is available?
> 4. As partisan materials are used, are students made aware of the nature and purposes of the organization or individual responsible for the material?
> 5. As partisan materials are used, are students helped to analyze persuasion techniques, faulty reasoning from evidence to conclusion, and other characteristics of this sort? (pp. 160-161)

Students should be reminded of their obligation to support their opinions with facts and reasoned interpretations; to listen thoughtfully to the views of others; and to defer judgment in the face of insufficient information or clearly conflicting values.

The teacher must also make clear to students the reasons for dealing with the issue. The objective is not to reach a consensus, but to analyze the issue from many angles and to assess the pros and cons of the available options. Practice with skills such as these should help students better understand and deal with similar issues in the future.

Peer Pressure

Gauld (1980) points out that a "teacher who is not willing to impose his own views on the students in the class may, nevertheless, unwittingly allow the coercive power of majority opinion to operate on 'highly sensitive teenagers who are particularly attuned to the judgements of others'" (Stewart 1975). Care should be taken to recognize such forces and to respect individual sensitivities and liberties.

In Conclusion

Education in the social and life sciences, as in other fields, is facing new territory as yet uncharted. That territory is circumscribed on one boundary by the rapid explosion of human knowledge and on the other by the inability of human systems and societies to respond rapidly in the realms of values, ethics, and public policy. Yet we may be sure that changes in knowledge and the ability of humankind to control the physical and social environment will result in a greater number of issues even more complex than those we face today. I believe that education must respond to this situation by creating within the curriculum opportunities for students to explore controversial topics. Moore (1979) concludes:

> It will probably take two generations to completely accomplish this task. Slowly a body of citizens and teachers who are more effective in the political process will emerge; this group will be better able to evaluate proposed solutions to social, economic, and medical problems and it will eschew the cheap, short-term gains that lead to long-term disasters. And, finally, they will provide and support a school system that teaches honestly and openly "what is" and use the information to plan for "what ought to be."

References

ALBURY, R. 1980. Ethics and experts: Some implications for science teaching. *The Australian Science Teachers Journal* 26(2):33-36.

BAKER, M. R., and DORAN, R. L. 1975. From an awareness of scientific data to concerns of mankind: Strategies for affective instruction in science. *Science Education* 59:539-558. As reported in Gauld, 1980.

BRANDENBURG, R. E. 1980. Memorandum decision. In Circuit Court, Eighth Judicial Circuit. *Lloyd Dale, Appellant vs. Board of Education*, Lemmon Independent School District, Lemmon, South Dakota.

BREASTED, M. 1970. *Oh! Sex Education*. New York: Praeger Publishers.

BIOLOGICAL SCIENCES CURRICULUM STUDY. 1978a. Guidelines for educational priorities and curricular innovations in the areas of human and medical genetics. *BSCS Journal* 1(1):19-29.

———. 1978b. Issues and controversy in teaching science. *BSCS Journal* 1(1): 33-37.

BUTTS, R. F. 1978. *Public education in the United States: From revolution to reform*. New York: Holt, Rinehart and Winston.

CALIFORNIA STATE DEPARTMENT OF EDUCATION. 1960. *Policies on teaching controversial issues adopted by California school districts*. Sacramento: The Department, n.d. p. 16 (Mimeo). As reported in NEA, 1963.

DOLL, R. C. 1974. *Curriculum improvement: Decision making and process*. 3rd ed. Boston: Allyn and Bacon, Inc.

EGGLESTON, J. 1977. *The sociology of the school curriculum*. London: Routledge and Kegan Paul.

FITZGERALD, F. 1979. Onward and upward with the arts. Rewriting American history. *The New Yorker*. I. February 26, 1979, pp. 41-77; II. March 5, 1979, pp. 40-91; III. March 12, 1979, pp. 48-106. As reported in Moore, 1979.

FREEMAN, J. 1978. Death and dying in three days? *Phi Delta Kappan* 60:118.

GAULD, C. F. 1980. Values education in the science classroom. *The Australian Science Teachers Journal* 26(2):5-11.

GOODLAD, J. I. and associates. 1979. *Curriculum inquiry: The study of curriculum practice*. New York: McGraw-Hill Book Company.

HICKMAN, F. M. 1979. Biology as a political science. *American Biology Teacher* 41:493-494.

HOOVER, K. H. 1976. *The professional teacher's handbook* 2nd ed. Boston: Allyn and Bacon, Inc.

KENNEDY, M. H. 1978. Contemporary biology issues. *In* Mayer, W. V. (ed.) *Biology teacher's handbook*. New York: John Wiley and Sons.

———. 1980. Biology education in the 1980s. Presented at the 8th Biennial Conference of the Asian Association of Biology Education, October.

MAYER, W. V. (ed.). 1978. *Biology teacher's handbook*. New York: John Wiley and Sons.

———. 1980. New needs—new programs. *BSCS Journal* 3(3)1-3.

MEIKLEJOHN, A. 1960. *Political freedom: The Constitutional powers of the people*. New York: Harper and Row. As reported in Butts, 1978.

MILLER, W. C. 1978. What will the future bring for educaton? *Phi Delta Kappan* 60:287-289.

MOORE, J. A. 1979. Dealing with controversy: A challenge to the universities. *American Biology Teacher* 41:544-547.

MUESSIG, R. A. (ed.) 1975. *Controversial issues in the social studies: A contemporary perspective*. Washington, D.C.: National Council for the Social Studies.

NATIONAL ASSOCIATION OF BIOLOGY TEACHERS. 1979. A look to the future. *NABT News and Views* 23(4):1-2.

____________. 1980. NABT position on scientific integrity. (Adopted by the Executive Committee, October 23, 1980.)

NATIONAL EDUCATION ASSOCIATION. 1963. *Deciding what to teach*. Washington, D.C.

ORGANIZATIONS FOR THE ESSENTIALS OF EDUCATION. 1978. *The essentials of education*.

PANEL ON SCHOOL SCIENCE. 1979. The state of school science on behalf of the National Academy of Sciences, National Research Council. *In* National Science Foundation, *What are the needs in precollege science, mathematics, and social science education?* Washington, D.C.

RESNICK, J. 1981. New emphases in biology. *National Association of Secondary School Principals Curriculum Report* 10(3).

ROPER, D. 1977. Parents as the natural enemy of the school system. *Phi Delta Kappan* 59:239-242.

STEWART, J. S. 1975. Clarifying values clarification: A critique. *Phi Delta Kappan* 56:684-688. As reported in Gauld, 1980.

Wieman v. Updegraff. 1952. Quoted in Bremner, R. H. (ed.) *Children and youth in America: A documentary history Vol. IV:* 1933-1973. Cambridge: Harvard University Press, 1974. As reported in Butts, 1978.

Selected Bibliography on Values Education

The following list is selected from the references provided in Gauld, C. F. (1980). Values education in the science classroom. *The Australian Science Teachers Journal* 26(2):5-11.

BAKER, M. R. and DORAN, R. L. 1975. From an awareness of scientific data to concerns of mankind: Strategies for affective instruction in science. *Science Education* 59:539-558.

BANKS, J. A. 1977. *Teaching strategies for the social studies*. Reading, Mass.: Addison-Wesley.

GENGE, B. A., and SANTOSUOSSO, J. J. 1974. Values clarification for ecology. *The Science Teacher* 41(2):37-39.

GENNARO, E. D., and GLENN, A. D. 1975. Science and social studies: An interdisciplinary approach to values and value decisions. *Science Education* 59:85-93.

GLUCK, P. G. 1977. Values clarification: The engineering of consensus. *Teachers College Record* 79:267-274.

HALL, W. 1973. *Science and decision-making*. London: Longmans and Penguin.

HARMIN, M., KIRSCHENBAUM, H., and SIMON, S. 1970. Teaching science with a focus on values. *The Science Teacher* 37(1):16-20.

HEPWORTH, A. J. 1978. Values education—some New South Wales experiences. *Journal of Moral Education* 8:193-201.

KUH, D. J. 1973. Value systems in life science instruction. *Science Education* 57:343-351.

LOCKWOOD, A. L. 1975. A critical view of values clarification. *Teachers College Record* 77:35-50.

__________. 1977a. Values education and the right to privacy. *Journal of Moral Education* 7:9-26.

__________. 1977b. What's wrong with values clarification. *Social Educaton* 41:399-401.

METCALF, L. E. (ed.) 1971. *Values education: Rationale, strategies, and procedures*. Washington, D.C.: Council for the Social Studies.

PIBURN, M. D. 1976. Values education: Insights from the social sciences. *The Science Teacher* 43(3):18-20.

RATHS, L., HARMIN, M., and SIMON, S. 1966. *Values and teaching*. Columbus, Ohio: Merrill.

SHATTUCK, J. B. 1970. Using the sciences for value clarification. *Science Education* 54:9-11.

SIMON, S., HOWE, L., and KIRSCHENBAUM, H. 1972. *Values clarification: A handbook of practical strategies for teachers and students*. New York: Hart.

SNYDER, P. J. 1976. Talking openly about sex. *The Science Teacher* 43(1): 28-29.

STAHL, R. J. 1979. Working with values and moral issues in content-centered science classrooms. *Science Education* 63:183-194.

STEWART, J. S. 1975. Clarifying values clarification: A critique. *Phi Delta Kappan* 56:684-688.

SUPERKA, D. P., *et al.* 1976. *Values education sourcebook*. Boulder, Colorado: Social Science Education Consortium.

Commentaries

The ever-present timidity of textbook publishers, the presence of power groups with vested interests, and the danger of inculcation by teachers are not insurmountable problems. I commend those groups who recommend that students have the right to study controversial issues, to have access to relevant information free from bias and prejudice, and to form and express their own opinions without threat of reprisal. The NEA has made some impressive contributions to the discussion through clear definition of a controversial issue and through its stand on rational discussion of controversial issues. Important steps are identification of relevant informaton, critical analysis, the formation of independent, supportable judgments, and a readiness to review the issue as times change.

The role of the teacher is critical. Should we explore more intensively the need for increased training in these skills while potential teachers are undergraduates and in practice teaching? Given the busy schedule of teachers today, it is unrealistic to expect them to go to the trouble outlined by Freeman (1978); but it is practical that school boards adopt some overall guidelines. It is probably possible to make more efficient use of classroom time to make room for discussion of controversial issues. Perhaps the reasons "elders" cannot find solutions may be that they were never given the opportunity to explore and openly discuss these issues in school. As Hickman (1979) says, developing decision-making skills will give students the self-confidence needed to make decisions in the future.

Hickman's paper is an objective, balanced, and thorough exploration of the benefits and risks of discussing controversial issues in the classroom. We need to encourage active, informed, rational, potentially sophisticated students who will, we hope, become enlightened adults equipped with decision-making skills and the ability to enjoy productive, meaningful lives.

Jane Abbott

Hickman's paper is a clarification of the problems and issues every teacher eventually confronts in dealing with controversial topics. This paper is filled with noble goals and ambitious philosophies about the role of the teacher as the promoter of "freedom of thought" in a democracy for "education for informed citizenship and enlightened decision-making."

She alludes, however, to the major problem that many public school educators face today: survival. Each of us graduated with a teaching certificate neatly tucked under our arm and with a robust assurance that what we were about to do was a noble, worthwhile endeavor—a "calling" of magnanimous promise. We had the confidence that what we could do with our students could make a difference. We have now been backed into the corner of survival. Given the realities of the present choices, which will the average public school teacher choose? The noble vision or the ignoble, but necessary, choices of survival?

This paper is a great help in clarifying the issue; it views the "controversy about controversy" from both philosophical and practical perspectives with clear advice about how to proceed once the decision is made to break the barriers a teacher must face in his or her school district. Controversies are virtually unavoidable in the biology teaching of the 1980s. We each will have to make a decision to teach or not to teach.

Roy K. Baer

Hickman's open and objective coverage of the pros and cons of including controversy in the high school curriculum is a realistic approach to the problem. The pressures and politics involved can be neither over-emphasized nor overlooked for all the reasons she explores. Implications for the classroom are focused upon the teacher. While textbooks must remain neutral in order to gain adoption, the burden of the responsibility to teach or not to teach controversial issues remains with the teacher.

Hickman suggests soliciting support from parents, boards of education, and administrators. That certainly would provide a sense of security for the teacher dealing with controversy. However, I cannot envision such groups being able to reach agreement among themselves. It is certain that the vote in many sectors would be negative. I would much prefer to approach such topics independently and in such a manner as to avoid inciting feelings that might end in disaster. Admittedly this requires skill and insight in directing both student/student and student/teacher interaction. This skill may come from experience or from special training in leading discussions. The development of classroom materials relating to controversial issues is necessary and could provide invaluable resources for such teaching.

Betty W. Long

Teachers cannot afford to shortchange young people by sidestepping controversial issues. Hickman has given valuable guidelines for teachers as they prepare to deal with issues that may present controversy. She also leads us into excellent literature through which further help is available. Careful groundwork to be done by teachers includes drawing up objectives, consulting with administrators, colleagues, parents and others. To be discussed and decided upon are such topics as scope of issues to be explored with students, techniques of discussion, sources and availability of information, evaluation of climates in the home and in local and larger communities. Finally, a set of guidelines, to be periodically reevaluated, needs to be drawn up and approved by administration in the school and district.

Sister Marian Catherine McGrann

My early decision to *discuss with* my students the social and other implications of biological knowledge and technology has rewarded me immensely. However, it was not without risk, and I made many errors in judgment, strategy, and topic selection. I probably learned more from my students when dealing with controversial issues than they learned from me. Over the years many of my former students have returned to tell me of their appreciation for their seminar discussions and other activities in my classes.

Hickman's sensitive treatment of whether to deal with controversial issues is one of the strongest parts of her paper. Some teachers, feeling fearful or inadequate about these matters, need encouragement and assistance if they are to move gradually into controversial issues. I wish her comments had been available when I began teaching biological science twenty-three years ago.

James A. Meyer

9

BIOLOGY EDUCATION: BEYOND THE STATUS QUO

Rodger W. Bybee
Jane Butler Kahle

The status quo is neither ethically nor legally defensible. We have an obligation to move beyond the status quo.

Is biology education meeting the challenges of scientific, social, and educational priorities? Unfortunately, the answer is NO. Currently, there are many indications of discrepancies between what is happening in biology education on the one hand and the new demands of science, society, and education on the other. This assertion is based, in part, on evidence in an earlier paper entitled "Biology Education in the Secondary Schools of the United States" (Hurd, *et al.*, 1980).

The concerns that biology teachers face daily—enrollments, budgets, facilities, and so on—are related to the larger social problems. Going beyond the status quo in biology education requires that we first understand the present situation. The first part of this paper describes the status quo, which itself provides evidence for needed changes. The second part outlines recommendations for going beyond the status quo in some crucial areas of biology education.

IN THIS PAPER we shall look at what we are teaching now and what we propose to teach in the future. We will examine and analyze thirteen areas—from the basics movement to the retention of women—in biology education. The information presented in the next sections is from three National Science Foundation surveys on the status of science education (Helgeson, Blosser, and Howe 1977; Stake and Easley 1978; Weiss 1978) and the 1976-77 National Assessment of Educational Progress. These reports represent the most comprehensive survey of science education ever completed.

The Status Quo—Circa 1978

Basics Movement

The exact impact of the "back to basics" movement on science programs is, at best, unclear. The assumed shift away from science-related subjects toward reading, writing, and computation is not perceived by teachers. Weiss (1978) reported that only 7 percent of science teachers rated "belief that this subject is less important than other subjects" as a serious problem. Thirty-nine percent thought that it was somewhat of a problem, and 51 percent indicated it was not a significant problem. In many cases it seems that existing science programs meet the criteria of "basics" since a review of the goals, curriculum, and instruction in biology (Hurd, *et al.*, 1980) reveals that biology education is not innovative, inquiry oriented, or socially relevant. It is basic biology! Perhaps we need to ask the next questions: How is biology basic in today's world? What biology is basic to resolving present problems? What biology is basic for the individual, the community, and the environment?

Budgets

Funding is perceived as a major problem facing science teachers. An estimated 66 percent of science teachers indicated funding is, to some degree, a problem (Stake and Easley 1978). It is evident that there is a need for improving the financing of science education. First, however, we need to recognize that we have just been through a period when science programs received extraordinary support. Second, there are funds, possibly more than we realize; but they are not available for the programs, institutes, and materials associated with the curriculum movement of the 1960s and 1970s.

Career Education

The recent emphasis on career education is being felt in science education. The impact is, in part, a response to public opinion. Public interest in career education will continue as long as the economy is unstable and unemployment is high. During a period such as this, individuals are concerned about jobs. Traditionally they turn to the schools for help in the form of career information and vocational training. When teachers, parents, and science coordinators were asked about vocational goals of science courses, they agreed that science courses should be more vocationally oriented; yet the majority would select a good general education over a vocational program if they had to select between them (Stake and Easley 1978).

There is evidence that some teachers tie science to vocations in industry and agriculture. In addition, the reduction of knowledge into smaller pieces, such as facts, skills, and concepts is thought to help with careers. Biology teachers can easily include information about many science-related careers, which go beyond the usual laboratory scientist, forester, and medical technologist. It should be pointed out also, however, that some science teachers question the vocational/career goal (Stake and Easley 1978).

It is instructive to compare the views of adults, as stated above, concerning career education with the views of students. Here are some results from the 1976-77 National Assessment of Educational Progress surveys of 17-year-olds: When students were asked: (a) "Would you like to work on a job that allows you to use science?," 39 percent indicated "yes" and 35 percent "no"; (b) "Would you like to visit a scientist at work?," 71 percent said "yes" and 16 percent "no"; (c) "Are there science-related jobs you would like to do?," 77 percent said "yes" and 9 percent "no"; (d) "Would you like to know more about jobs in science and engineering?," over 50 percent indicated that they would, while 39 percent indicated they would not; (e) "Would you like to work in a laboratory?," "yes" was the response of 43 percent of the students; 36 percent said "no"; (f) "Do you think working in a science-related field would be boring?," 46 percent of 17-year-olds think scientific work would *not* be boring, 31 percent think it would. Clearly, students are concerned and interested in information about careers in science.

Competency and Accountability

Concerns about accountability in education and competency-based science programs have increased in the last decade. State departments of education are the foci of the pressure for programs of this nature. To date, minimum competencies have had little impact on secondary sciences (Helgeson, Blosser, and Howe 1977). While 73 percent of states are planning competency programs, the implementation dates tend to be for some time in the future or not yet determined (Weiss 1978). By most indications the crest of this movement has passed, and education has survived yet another attempt to solve all of its problems with one simple remedy.

Curriculum

A number of NSF curriculum programs have been too difficult for average and below-average students. The difficulty is attributed to high reading levels and to the complexity of the scientific concepts included (Helgeson, Blosser, and Howe 1977).

Another curricular problem is that state regulations now require a broader range of science courses. Many of the new courses are related to societal issues such as energy or ecology. However, state regulations have reduced the amount of science required for graduation. They have also indirectly reduced money for science programs and, in some cases, eliminated programs (Helgeson, Blosser, and Howe 1977). On the whole, science teachers' reactions to regulation of the curriculum by state departments of education are mixed. The main problems are regulations that require change in the curriculum without providing time, money, or resources to facilitate that change.

The status of science curricula is further confused by science-related social issues. Such issues were largely neglected in the NSF science improvement curricula of the 1960s. Response to the concern for social issues is represented best by the rapid expansion of earth science programs between 1965 and 1975 and the implementation of environmentally and ecologically oriented courses in the biological sciences during the same time period. We think this trend is much more significant than other issues such as accountability and competency-based programs.

Equality of Educational Opportunity

The issue of equality of educational opportunity has resulted in a conflict between educational practices and democratic principles. To date, it seems the confrontation has been subtle and beneficial to neither teachers nor students (Stake and Easley 1978). The confrontation centers on the Fourteenth Amendment rights of students and the pedogogic practices of teachers. Since *Brown v. Board of Education*, 1954, there has been more legislation and regulation requiring that handicapped, minorities, children from low income families, and women not be denied their right to an education. Some of these issues are discussed at length in a later section.

The impact of these requirements on science education is not clear. Science teachers are apparently in a state of dissonance between their beliefs in human rights and the concrete need to modify their teaching styles. Our position is clear and unequivocal: the status quo is neither ethically nor legally defensible. We have an obligation to move beyond the status quo.

A concrete, but less direct, result of legislation for equality of educational opportunity is the reduction of science budgets so that funds can be allocated for other purposes. The reduction of funds subsequently influences instructional materials, laboratory equipment, field trips, and inservice education (Helgeson, Blosser, and Howe 1977). While this result is unfortunate, there is a greater harm done through discrimination than by the reduction of revenues. Creative biology teachers have always been able to find ways to work around reduced budgets. Some of this creativity will undoubtedly be required in the future.

Enrollments

Enrollments in secondary schools increased during the 1950s, 1960s, and 1970s but are now declining. Principals surveyed in the NSF studies rank declining enrollments fourth among all problems in science and mathematics education. However, science teachers did not rank declining enrollments among their most important problems (Weiss 1978).

For science education, the most obvious result of the decline in enrollment is a reduction in revenues. Reduction in enrollments will directly affect science programs through the reduction in budgets, facilities, courses, and staff. Some have argued there is not an oversupply of teachers, but an undersupply of money. Federal support of science education has declined since the late 1960s, as have state and local support (Helgeson, Blosser, and Howe 1977).

Facilities and Equipment

Science equipment and supplies, or their perceived lack, greatly influence science teaching. Funds for supplies are, obviously, a major factor in science programs. In other sections it has been pointed out that budgeting restraints underlie several other problems in science education. In light of these statements, 26 percent of science teachers rate inadequacy of facilities as a serious problem, and 42 percent indicate that it is somewhat of a problem. To some degree, facilities are a problem for 68 percent of science teachers (Weiss 1978).

Instruction

Thirteen percent of secondary science teachers are currently teaching at least one class for which they feel inadequately qualified. Most of these teachers are referring to courses within science (Weiss 1978). When asked about the areas of greatest need, science teachers list several items directly related to instruction. Approximately 66 percent of all teachers surveyed (science, mathematics, and social studies) indicate that they need information about instructional media. Sixty-one percent of the teachers polled reported a need for assistance in learning new teaching methods. Most science teachers surveyed felt that they were competent enough to implement the discovery/inquiry approach. However, the evidence indicates they did not do so (Stake and Easley 1978). Forty-eight percent of all science, mathematics, and social studies teachers indicate they would like assistance in the use of manipulative materials. This need may relate to the fact that manipulative materials are generally used less than once a week in more than 50 percent of all science, mathematics, and social studies classes (Stake and Easley 1978). One other instructional problem which science teachers rated high was the lack of materials for individualized instruction. Twenty-nine percent of teachers rated this a serious problem and 39 percent somewhat of a problem (Weiss 1978).

Students and Classrooms

Disruptive students and classroom management teachniques are concerns of science teachers (Helgeson, Blosser, and Howe 1977). In the NSF survey conducted by Weiss (1978) science teachers were asked about the difficulty of maintaining discipline. Six percent of junior high (grades 6-9) science teachers indicated discipline was a serious problem; 30 percent indicated it was somewhat of a problem; and 64 percent indicated discipline was not a significant problem. Nine percent of high school (grades 10-12) science teachers indicated discipline was a serious problem; 31 percent indicated it was somewhat of a problem; and 58 percent reported discipline was not a significant problem. In total, 36 percent of junior high and 40 percent of senior high science teachers reported that maintaining discipline in the classroom was a problem.

These data are consistent with a national survey conducted by the National Institute of Education, *Violent Schools-Safe Schools* (NIE 1978). Approximately 8-10 percent of schools report violence and disruption as a serious problem. These data, along with public opinion polls (Gallup 1979), indicated that discipline problems and the resolution of conflicts with students are among the dominant problems in American classrooms.

Despite these statistics, surprisingly small percentages of science teachers indicated a need for assistance in maintaining discipline. Seventy-nine percent of science teachers in grades 7-9 and 83 percent of science teachers in grades 10-12 indicated that they did *not* need assistance. Nine percent of science teachers at both levels, 7-9 and 10-12, reported that they receive adequate assistance in maintaining discipline. Ten percent of junior high and 5 percent of senior high science teachers did not receive adequate assistance with discipline problems. In all, approximately 36 to 40 percent of science teachers reported discipline as a problem. But very small percentages indicated they receive no assistance.

Teacher Training

The training of biology teachers is a two-part process: first, preservice or undergraduate education which includes both general education and professional training and, second, inservice or postgraduate education. Usually, undergraduate courses taken by preservice biology teachers are the same courses taken by students preparing for professional or graduate schools. Large, impersonal lecture courses with structured laboratories are the most common format; and prospective teachers have few opportunities to participate in openended inquiry laboratories, to lead fruitful discussions, or to ask and respond to penetrating questions. The infrequent use of creative inquiry teaching in biology classrooms seems to be related to the fact that teachers rarely experience it in their college preparation.

The professional education component of preservice science teacher education is divided between instruction in elementary and/or secondary methods of teaching. Few institutions have programs specifically designed for middle school or junior high biology or life science teachers. Student teaching is most often a full-time experience for less than a semester.

Upon completion of an approved undergraduate teaching program, the preservice teacher receives a basic or provisional certificate effective for several years. A temporary certificate may be renewed or exchanged for a more permanent license after teaching several years and/or after accumulating additional academic credits. There is a great deal of flexibility and variety in requirements concerning the nature of this advanced work, not all of which needs to be part of a degree program or in the teacher's subject area. For example, over 75 percent of 7-12 science teachers today have taken more than ten hours of graduate work, and over 50 percent hold master's degrees. However, fewer than 40 percent have taken graduate courses in the science they are teaching (Weiss 1978).

In recent years, the NSF-funded summer and academic year programs have been popular inservice activities. For example, over 50 percent of science teachers in grades 7-12 have attended one or more NSF summer or academic year institutes. The teachers participating in NSF programs have been older, male, and primarily from larger schools in the Western and North Central regions of the country. After participation, teachers are more likely to use NSF curricular materials, to utilize the laboratory, and to stress student-centered activities. Furthermore, there is a consistent trend toward better student performance with increased teacher participation in NSF institutes (Helgeson, Blosser, and Howe 1977; Stake and Easley 1978; Weiss 1978). The federally supported summer and academic year institutes, then, have influenced the type of biology teaching which occurs in the classroom today.

The three national assessments of science achievement (NAEP

1977, 1979) have documented an overall decline in science scores (NAEP 1979). Statistically significant declines in achievement on test exercises have been noted between the first and second science assessments in both natural and physical science for all three age levels—9, 13, and 17. In 1977, a further decline was noted for 17-year-olds in both biological and physical sciences, although the decline was greater in the physical sciences. Nine- and 13-year-olds did not decline in achievement in biology exercises from the period 1973 to 1977, but both groups did decline steadily in achievement on the physical science questions (NAEP 1978c). This persistent decline, as well as the consistent lower achievement of significant groups of our students, is a critical problem in biology teaching. Generally, minorities, females, those living in disadvantaged urban areas, and those living in the Southeast perform consistently below national averages on all cognitive science items. For example, blacks, Hispanics, and disadvantaged urban students scored ten to twelve percentage points below the national averages in the most recent assessment. Likewise, females averaged 2.5 percentage points below the national means at ages 9, 13, and 17 (NAEP 1978c). In addition, within minority groups, female achievement falls below that of males. For example, Hispanic females are reported below Hispanic males in science and mathematics achievement at all age levels (NAEP 1977). For these segments of our population, then, the status quo means lower achievement in science classes, less understanding of scientific concepts, and decreased access to science-related careers.

Minorities

Factors which contribute to these lower achievement levels permeate current society, schools, and biology classrooms. For example, the National Research Council (1979) has found that although minorities in 1974 constituted almost 11 percent of the employed labor force, only about 5 percent hold jobs in science and engineering. Among the four generally identified minority groups, persons of Asian origin are overrepresented in the sciences and engineering, while the other three groups, American Indian/Alaskan native, blacks, and Hispanics are underrepresented. Of these three groups, blacks make up the most significant proportion of our population; they comprise 15 percent of the 18-21-year-olds and 10.7 percent of the total undergraduate population. However, blacks account for only 6.9 percent of undergraduates majoring in the biological sciences (Office of Civil Rights 1976). At the graduate school level, the number of minorities receiving doctorates in scientific disciplines is even lower. In 1977, they constituted less than 4 percent of Ph.D. recipients in all science and engineering fields. The lack of educational opportunities is reflected in the occupational data, described by Hale (1978). The study indicates that although blacks constitute 10 percent of the U.S. population, they account for only 1 percent of the physicians and engineers and 2 percent of the dentists.

The root of this underrepresentation may be traced to the early adolescent years. Kahle (1979) found that blacks are both formally and informally segregated from science and advanced mathematics classrooms. As she noted, "Classrooms are informally segregated, by the requirement of prerequisite classes. Many minorities do not take science courses because they have not been counseled into the academic 'tracks' or into the prerequisite mathematics classes" (p. 82). Rowe (1977) has also noted that ". . . middle-class blacks are choosing to enter politics, law and other social sciences, but not the research sciences or engineering. Even when blacks are successfully recruited as science students, their drop-out rate is high and they frequently ask to change profession enroute" (p. 34). The current state of biology education is, therefore, not meeting the needs of minority students.

It is evident that many gifted young people are undertrained and underutilized in the sciences and in mathematics. In 1979, the National Research Council noted that a large fraction of the nation's corporate executives and half of the federal decision-makers come from science and engineering backgrounds. The report goes on to state that as children approach adolescence, the availability of role models becomes an important factor in their selection of future careers. Mary Budd Rowe (1977) also cites this as a possible reason for blacks choosing fields other than the sciences. Studies have shown that although parents are listed by adolescents as the individuals most responsible for their career choices, associations with other adults holding specific occupations are second in importance (Pallone, Hurley, and Richard 1973). There is a need, therefore, to provide minorities with more role models by increasing the number of minorities teaching biology.

Women

The dearth of women in scientific or mathematical fields has been well documented. "As a percentage of Ph.D.'s, women have just now in the 1970s achieved the same level they enjoyed in the 1920s (Brown, Aldrich, and Hull 1979, p. 1). Another finding of this report is that in elementary and high school, girls and boys interested in science are treated differently by parents, teachers, and friends. "Girls found ambivalence, lack of encouragement, and messages that what they were doing was inappropriate, impractical or unacceptable. Boys encountered much wider acceptance of their intentions as appropriate and admirable, particularly in terms of future economic status and a successful career" (Brown, Aldrich, and Hull 1979, p. 1).

Women, especially minority women, are underrepresented in science and mathematics according to the National Research Council (1979). Although women now comprise approximately 50 percent of the work force, only 6 percent are employed as scientists and engineers. Again, one can look to the critical early adolescent years and the paucity of role models. Weiss (1978) has reported that although the vast majority of elementary teachers are female (96 percent), the ratio changes at grade seven, where only 41 percent are females. In high schools, approximately 24 percent of the science teachers are women, although we may safely assume that most of them are biology teachers since most science teachers are biology teachers. In addition to the lack of women role models, there is evidence that within science classes, girls are treated differently. The attitudes toward science component of the 1977 NAEP survey indicates that at age 17, 23 percent fewer girls than boys have wired an electrical circuit, 18 percent fewer girls have made a magnet, and 11 percent fewer have taken something apart to see how it works (NAEP 1978b). However, the same document reports that more girls than boys have cared for aquaria or plants, more socially acceptable female activities. Bowyer, Linn, and Stage (1980) report that differences in male and female achievement scores on the two recent National Assessment surveys in science and mathematics are directly proportional to the number of semester hours taken in science and mathematics courses. For example, on the whole, females take one-quarter of a semester less mathematics than males, and this fact alone accounts for the achievement differences.

The lack of minorities and women as role models is an acute problem in society, in our schools, and in our biology classrooms. This lack permeates the status quo and indicates one area where change must occur. Biology teachers are in a unique position to effect that change. The NSF surveys indicate that 80 percent of high school students take biology; and for 50 percent, it may be their only science course (Helgeson, Blosser, and Howe 1977; Stake and Easley 1978; Weiss 1978). Measures and materials designed to encourage women and minorities, therefore, must be incorporated into our biology classrooms in order to retain women and minorities in science-related fields.

The Status Quo—An Analysis

There are at least three levels at which one can view the problems described above. Analyzing and moving beyond the status quo requires consideration of practical problems, programs and policies, and purpose.

Practical Problems

The first level is everyday practice. We have addressed some of these problems: reduced budgets, declining enrollments, and inadequate facilities. Biology educators are starting to realize that logical solutions do not always resolve these problems. Requesting higher budgets, reallocating funds, and controlling spending may offer temporary relief but often have other unintended consequences. The reason that problems may continue is that they originate at deeper, more fundamental, levels. The problems will not disappear solely from action outside of biology classrooms. Symptomatic of this level are statements such as "if only we had more money," or "if NSF would only fund this proposal our problems would be solved." The resistance of first level problems to simple solutions requires that we continue the analysis. Let us look at the next level at which problems are described.

Programs and Policies

The second level has to do with programs and policies. Here the argument is that graduate programs, preservice and inservice teacher education programs, and the biology curriculum at all levels have basic defects that must be corrected. Examples we have discussed include: the basics movement, career education, curriculum, teacher training. The logical response is curriculum development of some kind. Competency-based programs, career awareness, science as a basic, and special topics such as energy, population, or ecology all serve as examples of the curriculum development response. Still, we conclude that curriculum innovation has not been the solution; and, while it is essential, it probably will not be *the* solution in the future.

At the same level there exists a host of regulations and policies outside of science education. Mainstreaming, accountability, competency, and equality of educational opportunities were discussed. There is no doubt that actions must be taken on issues such as mainstreaming and equality of opportunity for minorities and women. Still, one is hardly convinced that such actions will remedy the many problems of status quo in biology education. We must move to the third level of analysis.

Purpose

The third level at which we can analyze the status quo is that of purpose. We must ask of ourselves and of the profession—what are the goals we hope to achieve? What are the intended results of education in the biological sciences? What are the larger aims of science education? If we can reach some agreement on answers to these questions, then we will have a common purpose on which to build our programs, design our policies, and direct our practices. But first it is crucial that we recognize our condition. We suggest that our fundamental problem is that we need a new direction—a new purpose in biology education.

Suggesting that the fundamental problem of biology education is one of purpose is not to argue that all the problems should be ignored while we seek new goals. First, the problems cannot be ignored. Second, many of the problems are connected since we are discussing one system—biology education. Third, it would be naive to suggest that our problems would solve themselves if only we identified a new purpose.

Conversely, the possibility of resolving the myriad problems without identifying a new direction is small. The process of going beyond the status quo in biology education requires three steps: recognizing the need to change; establishing a clear direction and purpose; and developing appropriate programs, policies, and practices based on the new purpose. The problems described earlier signal that the old goals and aims no longer provide the purpose of biology education. Thus, it is safe to say we have passed the first step. We have perceived the need to change.

In the past we have tended to jump from the first step, the need to change, to the third step of developing programs and changing practices. While these changes are no doubt necessary, it is essential that we first define a clear direction and purpose. In the next section, some of the central aspects of a new direction and purpose are described.

Beyond the Status Quo

Our discussion of purpose may have left some saying, "sounds good, but what should our purpose be? How do we determine a new direction?" New directions in biology teaching will evolve from our history, our philosophy and, above all, our relationship with society. (For example, see Bybee 1977a,b.)

Society and Biology Education

Seeking a new direction for biology teaching starts with an understanding of our role in society. Consider the topics discussed in this volume: biology for life and living, ethics and values, public policy, genetics, human sexuality, and controversial issues. Without exception, these topics are related to human and social problems. We need not look far to find other biology-related social issues presented in popular books such as *The Eighth Day of Creation* (Judson 1979) or *The Wooing of Earth* (Dubós 1980). But our programs do not really represent these current biological issues.

A review of the NSF surveys (Helgeson, Blosser, and Howe 1977; Stake and Easley 1978; Weiss 1978) showed that knowledge continues to be the emphasis of biology programs. This knowledge is aligned with the structure of biology as a discipline and *not* with the social issues just described. Inquiry and problem solving, two major goals of the 1960s, are seldom seen as contemporary goals. Careers in science are presented but not substantially. The most glaring omissions in biology programs are topics related to personal needs and to social issues (Hurd, *et al.*, 1980).

Evidence that change is in progress is seen in public attitudes toward science-related social issues, confidence in science, and support for scientific research. Presently there is some ambivalence on these issues and the trend seems to be toward one of doubt, rather than belief (Bybee, *et al.*, 1980). There is also evidence that students are aware of many science-related social issues, but they lack conceptual understanding of the problems and of their role in resolving them (Harms, Bybee, and Yager 1979). While we cannot expect biology teachers to eliminate this situation, it is reasonable to expect them to set a new direction that can contribute to the resolution of such problems.

A New Direction for Biology Teaching

Few dispute the need to change. But, is the direction of change clear? We think it is much clearer than most imagine. The evidence strongly suggests that we should be directing our programs toward greater recognition of personal and social issues related to biology. In an earlier paper Bybee (1979b) suggested:

> In the next decade policies for science education programs should include appropriate cognitive, affective, psychomotor, and social objectives to: (1) fulfill basic human needs and facilitate personal development; (2) maintain and improve the physical and human environment; (3) conserve and efficiently use natural resources; (4) develop greater community at the local, regional, national and global levels. (p. 159)

The components of this aim should be familiar to biology teachers. They are the basic units of ecological study—organisms, communities, and the environment. But because *human* ecology (Miller 1979) is the central concern, modifications of the basic units are needed. The human ecological goal implies new directions if we are to move beyond the status quo.

Curriculum

The basic difference between status quo biology programs and new programs is context. The basic concepts of biology such as heredity, evolution, nutrition, behavior, continuity, structure and function, diversity, integration life cycles, and energetics will remain. In the 1960s these important concepts were taught in an abstract context—the structure of biology as a discipline. Indeed, there is nothing wrong with this type of organization of scientific knowledge and processes. What we are suggesting is the presentation of the same basic concepts in a context that is personally meaningful and socially relevant. In table 1 we have used heredity to demonstrate contextual differences in curriculum organization.

TABLE 1. Concept: Heredity

Context: Structure of the Discipline	*Context: Personal and Social*
Mendel's Laws	Human Abnormalities
Chromosome Theory	Sex Linkage
Probability	Genetic Disease
	Eugenics
	Genetic Engineering

Some say that the personal and social themes can be absorbed by the old model; or, that students need to know Mendel's laws, chromosome theory, and probability before they study human heredity. There is little, if any, evidence supporting the latter position. We wish to make clear that we are not eliminating "basic" genetics. Mendel's laws, chromosome theory, and probability all apply to contemporary hereditary issues. Why not teach those concepts in that context? Students will be making many more decisions about hereditary issues in the right column of table 1 than the left one.

Curriculum materials are available with the orientation we have described. For example, the human ecological goal is well developed in such BSCS programs as *Investigating Your Environment* (1975), *Energy and Society* (1977), *Land Use* (1979), and *Quality of Life and the Future* (1980). These materials can be used individually as supplements to a biology program, or, they can be the semester or year-long biology program. The BSCS is also preparing instructional materials in genetics closely aligned with the personal and social context suggested in table 1.

Instruction

As biology students and teachers become more diversified, instructional techniques and classroom practices will change also. Generally, teaching facilities and instructional practices in the future will have to extend beyond the classroom to the total school and the community. Biology teachers will need to use the varied resources of their community, provide choices of activities, utilize a range of reference and resource materials, and cooperate with other teachers, both in the sciences and in the other subjects, especially social sciences. All of these suggestions can also help reduce budgetary and enrollment problems.

One specific technology which biology teachers may use to move beyond the status quo is the microcomputer. It is anticipated that soon microcomputer technology will be incorporated into the lives of every citizen. The computer industry predicts that by 1985 approximately six million households will have home computers. In order to assure that all students have equal opportunities for using this technology and learning the prerequisite skills, teachers must incorporate microcomputers into their curriculum. "Unless the school system provides basic computer skills, the children of affluent families will be the ones who have access to home computers and acquire skills that increase their education advantage" (NSF 1979, p. 24).

Microcomputers may be used to manage classes (*e.g.* grades and attendance), review, test, teach problem-solving skills, present simulations, and store and compare data. Once students are accustomed to the use and capabilities of the microcomputer, they may be able to use it in other content areas and in non-traditional educational settings such as museums and homes. The creativity afforded the student, utilizing a microcomputer within or outside the classroom, is tremendous. It is possible to simulate many scientific experiments and to allow unique approaches to data analysis. Microcomputers make possible individual experiences with problems that otherwise may not be encountered by students.

Some of the projected learning activities that will involve new instructional approaches are as follows:

1. Development of classes into model social units.
2. Use of a variety of media, instructional techniques, and educa-

tional resources (simulation games, data gathering techniques, decision-making experiences).

3. Cooperative projects with community agencies (Girl and Boy Scouts, Planned Parenthood, Cancer Society, and so on).

4. Data gathering and survey analyses of biosocial issues and problems.

5. Provision of information concerning science-related careers through media, role models, and resource people.

These experiences indicate a teaching situation which is very different from the self-contained classroom described in the three NSF surveys. Biology teaching will require interdisciplinary skills and multidisciplinary information in a flexible, open classroom-laboratory which extends beyond the schoolhouse door and into the surrounding community.

Handicapped Students in Science and Science Education

Public Law 94-142—the Education for All Handicapped Children Act of 1975—requires that special students be integrated into the biology classroom to the maximum extent possible. There is every reason to believe special students can and should learn biology. In many cases the transition of special students into the biology classroom can be smooth. Two factors contribute to this transition: (1) the biology teacher's acceptance of the importance of biology in the life of special students and (2) the teacher's and students' understanding of the unique needs of any special students in biology classes. Space does not permit a discussion of the various ways science teachers can assist special students. Bybee (1979c) outlines specific suggestions for different populations of special students.

Beyond the biology classroom handicapped students may find productive careers in the biological sciences. Talent comes in many forms and we need the best and brightest in biology—regardless of their handicaps, color, or gender.

Minorities in Science and Science Education

In spite of lower achievement results, analysis of the responses of black students to the attitudinal items in the 1977 science assessment showed positive attitudes toward science, scientists, science research, science education, and societal problems attributed to science (NAEP 1978b). For example, fewer black students than white students found science boring; more blacks found it fun; and 48 percent of blacks (compared to 35 percent of whites) liked to go to science classes. Furthermore, 82 percent of black students (58 percent of white 17-year-olds) thought that science should be required, and 73 percent of black 17-year-olds (54 percent of white 17-year-olds) considered science education worth the necessary effort and expense.

These attitudes are interesting in light of other responses indicating that black students and those in urban areas had fewer science experiences both in and out of school. For example, approximately 20 percent fewer black students had seen a fossil, the North Star, a sprouting seed, or the moon through a telescope; approximately 10 percent fewer black students had visited a planetarium, a zoo, or a museum.

The 1976-77 National Assessment survey of attitudes also included questions concerning classroom practices and procedures. Responses to these questions help explain the positive attitudes of minorities toward science. They also provide directions for changing the status quo of minorities in biology. In all categories that indicated more individualized modes and materials of instruction blacks and students from urban disadvantaged areas had more experiences. That is, more minority students had selected the sequence of topics, had decided when to take a quiz, had completed a science project, and had gotten to work at their own speed (NAEP 1978b). The national concern with compensatory education for culturally and economically deprived minorities may have produced this pattern.

Research has carefully documented both the affective and achievement gains that accrue from utilizing individualized instruction with disadvantaged students (Kahle, *et al.*, 1977a,b). The evidence of greater utilization of individualized materials and/or modes of instruction in urban areas is one of the positive trends indicated by the NAEP survey. Such classroom procedures, which foster positive self-concepts by encouraging achievement, may have positively affected the attitudinal responses of minority, disadvantaged students. If this is the case, then, more individualized biology teaching methods may help us move beyond the status quo to improve both the achievement levels and the retention rates of minority students in biology.

Women in Science and Science Education

The factors affecting numbers and levels of women in science classes and in scientific careers are basically social, not economic. Girls are subtly and unconsciously discouraged from enrollment and achievement in science and mathematics classrooms (Erlick and LeBold 1977; Jacobs 1978). In the future, girls may be encouraged by extensive use of women role models in biology classes and in counseling activities. Profiles of women biologists and teachers must be incorporated into curricular materials, and media depicting unconscious sex bias must be made available for teacher training activities (Fennema 1980).

Specific instructional and managerial activities designed to enhance both the participatory and achievement levels of girls in biology classes must be effected. For example, laboratory and workbook activities that provide spatial experiences may enhance the spatial abilities of adolescent females. Laboratory groups must be carefully structured so that girls perform laboratory activities. The NAEP data, for example, show that at age 9 girls have had as many 'hands-on' science experiences as boys, but the percentage of their activities falls off dramatically at age 13 (NAEP 1978b).

Generally, instructional strategies must meet the needs of learners who have been discouraged under present conditions. Compensatory instruction—suitable for different ability levels, types of logical reasoning, cognitive styles, perceptions of fate control, and spatial abilities—are needed to move beyond the status quo. Our technologically oriented and scientifically based society can no longer underutilize the skills and abilities of over half the population.

Teacher Education in Science

To meet the needs of our scientific and technological society, preservice and inservice education needs to focus on the complex interactions between biology and society as well as on the meaning and nature of learning. Preservice training should include courses in the natural and physical sciences that stress inquiry, data collection, and holistic strategies of investigation. In addition, the preservice training of the biology teacher should include more work in the social sciences. Courses in geography, economics, sociology, and anthropology have both content and research techniques which are applicable to studying science-related social or cultural problems.

In the professional education component of preservice preparation, skills in the human-relations aspects of biology teaching such as listening, responding, discussing, and writing will be emphasized in addition to technical skills of planning, organizing, assigning, and evaluating. Courses specifically devoted to the teaching and learning of early adolescents and adolescents will need to be included. In the future, there is a need for both professional education courses and certification programs specifically designed for middle school and junior high school (grades 5-9) life science teachers. Efforts must be made to make teacher preparation and certification a continuing process, based upon both subject matter knowledge and teaching competency.

A recent comprehensive report by the National Science Foundation identified six new areas in which biology teachers must receive preservice training in order to meet the needs of their students. All of these indicate bioethical and biosocial teaching as a future direction:

1. Decision-making and the concept of risk,
2. Nature of proof,
3. Values, preferences, and prejudice in thinking,
4. Relation of science and mathematics to the real world of the student,
5. Examples of all types of scientific and mathematical careers,
6. Examples of interdisciplinary connections and research between science and mathematics (NSF 1978).

Preservice biology teacher training needs to address the cognitive, personal, and societal needs of 10- to 15-year-olds in order to promote changes. Transitional learning skills that bridge the gap from general, elementary schools to content specific, secondary schools are necessary. Biology methods courses must emphasize the use of inquiry problems, data collection, and analysis using techniques from the social sciences, holistic strategies of investigations, and integration of mathematics and science. Specific examples of the contributions of women and minorities must be presented; and community and out-of-school resources need to be employed. Preservice training that includes all of these elements should result in teachers able to teach diversified topics and students in the step beyond the status quo.

Inservice training activities in the future must respond to the public's demand for "competency." For example, the most recent Gallup Poll (1979) of the Public's Attitudes Toward the Public Schools indicated that 85 percent of the public agree that teachers should "be required to pass a state board examination to prove their knowledge in the subject(s) they will teach." Furthermore, 85 percent also think that such exams should be passed periodically to indicate if teachers are staying abreast of their field(s). In the future, inservice education must stress continued professional growth as opposed to intermittent presentations of new materials, approaches, or techniques. Specifically, two areas need to be stressed. First, inservice programs for biology teachers should be interdisciplinary; that is, attention should be given to social issues and information from other disciplines such as economics, sociology, and political science. Furthermore, ethics and values must be incorporated into these programs. Second, inservice programs should be conducted to orient biology teachers in the use of out-of-school resources. Inservice workshops and institutes must involve teachers of all sciences and social sciences at the local level with institutions and agencies at the state and national levels. Furthermore, there must be coordination and cooperation across these levels.

Summary and Conclusion

We assert that biology education is not meeting the challenges of the 1980s. Evidence from national surveys sponsored by NSF and NAEP support that contention. Analysis of the status quo reveals a deep problem of purpose for biology education. Purpose is aligned with the larger aims and goals underlying practice, programs, and policies. Now, the need is to move beyond the status quo to establish a new direction in biology teaching. In order to move beyond the status quo, biology teachers must first clearly recognize the role of biology and biology education in our society. New directions have been identified. In the broadest sense, human ecology is the new context for biology practices, programs, and policies. Human ecology as a new theme has important implications for some critical areas such as curriculum, instruction, special segments of our students, and teacher education. The message is loud and the direction clear. It is time to move beyond the status quo in biology education.

References

BIOLOGICAL SCIENCES CURRICULUM STUDY. 1975. *Investigating your environment.* Reading, Mass.: Addison Wesley Publishing Company.

____________. 1977. *Energy and society*. Northbrook, Ill.: Hubbard.

____________. 1979. *Land use*. Boulder: BSCS.

____________. 1980. *Quality of life and the future.* Boulder: BSCS.

BOWYER, J. B., LINN, M. C., and STAGE, E. K. 1980. *Issues in science education: Women in science*. Paper presented at the annual meeting of the National Association for Research in Science Teaching, April 1980, Boston.

BROWN, J. W., ALDRICH, M. L., and HULL, P. Q. 1979. The participation of women in scientific research. *Comment* 11:1-3.

BYBEE, R. 1977a. The new transformation of science education. *Science Education* 61(1):85.

____________. 1977b. Toward a third century of science education. *American Biology Teacher* 39(6):338.

____________. 1979a. Science education and the emerging ecological society. *Science Education* 63(1):95.

____________. 1979b. Science education for an ecological society. *American Biology Teacher* 41(3).

____________. 1979c. Helping the special student fit in. *The Science Teacher* 46(7).

BYBEE, R., HARMS, N., WARD, B., and YAGER, R. 1980. Science, society and science education. *Science Education* 64(3).

DUBÓS, R. 1980. *The wooing of earth*. New York: Charles Scribner's Sons.

ERLICK, A. C., and LEBOLD, W. K. 1977. *Factors influencing the science career plans of women and minorities 1977. Final Report*. West Lafayette, Ind.: Purdue University, Department of Freshman Engineering and the Measurement Research Center.

GALLUP, G. 1979. The eleventh annual Gallup poll of the public's attitudes toward the public schools. *Phi Delta Kappan* 61:33-45.

HALE, F. W., JR. 1978. An agenda for excellence: You can make the difference. *Phi Delta Kappan* 60:204s-206s.

HARMS, N., BYBEE, R., and YAGER, R. 1979. Science and society: A review of the national assessment data for 13- and 17-year-olds with implications for policies and research. *In* National Assessment of Educational Progress, *Attitudes Toward Science* (NAEP Science Report No. 08-S-2). Denver: Education Commission of the States.

HELGESON, S. L., BLOSSER, P. E., and HOWE, R. W. 1977. *The status of pre-college science, mathematics, and social science education: 1955-1975*. Volume I: Science education. Washington, D.C.: U.S. Government Printing Office.

HURD, P. D., BYBEE, R. W., KAHLE, J. B., and YAGER, R. 1980. Biology education in secondary schools of the United States. *American Biology Teacher* 42(7):388.

JACOBS, J. E. (ed.) 1978. *Perspectives on women and mathematics*. Columbus,

Ohio: ERIC Clearinghouse for Science, Mathematics and Environmental Education.

JUDSON, H. F. 1979. *The eighth day of creation*. New York: Simon and Schuster.

KAHLE, J. B. 1979. An analysis of minority attitudes toward science from the 1976-77 National Assessment of Science Results. *In* National Assessment of Educational Progress, *Attitudes Toward Science* (NAEP Science Report No. 08-S-2). Denver: Education Commission of the States.

———, NORDLAND, F. H., and DOUGLASS, C. B. 1976a. An analysis of an alternative instructional model for disadvantaged students. *Science Education* 60:237-243.

———, DOUGLASS, C. B., and NORDLAND, F. H. 1976b. An analysis of learner efficiency when individualized and group instructional formats are utilized with disadvantaged students. *Science Education* 60:245-250.

MILLER, G. T. 1979. *Living in the environment*. Belmont, Calif.: Wadsworth Publishing Company.

NATIONAL ASSESSMENT OF EDUCATIONAL PROGRESS. 1977. *Hispanic student achievement in five learning areas: 1971-1975* (NAEP Science Report No. BR-2). Denver: Education Commission of the States.

———. 1978a. *Science achievement in the schools: A summary of results from the 1976-1977 National Assessment of Science* (NAEP Science Report No. 08-S-01). Denver: Education Commission of the States.

———. 1978b. *The third assessment of science, 1976-1977: Released exercise set* (NAEP Science Report No. 08-S-21). Denver: Education Commission of the States.

———. 1979. *Three national assessments of science: Changes in achievement, 1969-1977* (NAEP Science Report No. 08-S-00). Denver: Education Commission of the States.

NATIONAL INSTITUTE OF EDUCATION. 1978. *Violent schools–safe schools*. Washington, D.C.: U.S. Government Printing Office.

NATIONAL RESEARCH COUNCIL. 1979. *The state of school science: A review of the teaching of mathematics, science, social studies in American schools and recommendations for improvements*. Washington, D.C.: National Research Council.

NATIONAL SCIENCE FOUNDATION. 1978. *Early adolescent: Perspectives and recommendations* (SE-78-75). Washington, D.C.: NSF.

———. 1979. *Technology in science education: The next ten years* (SE-79-57). Washington, D.C.: NSF.

OFFICE OF CIVIL RIGHTS. 1976. *Racial and enrollment data from institutions of higher education*. Washington, D.C.: U.S. Government Printing Office.

PALLONE, N., HURLEY, R., and RICHARD, F. 1973. Further data on key influence of occupational expectations among minority youth. *Journal of Counseling Psychology* 20:484-486.

ROWE, M. B. 1977. The forum: Why don't blacks pick science? *The Science Teacher* 44:34-35.

STAKE, R. E., EASLEY, J., *et al*. 1978. *Case studies in science education*. Volume

I: The case reports; Volume II: Design, overview and general findings. Washington, D.C.: U.S. Government Printing Office.
WEISS, I. R. 1978. *Report of the 1977 national survey of science, mathematics, and social studies education*. Washington, D.C.: U.S. Government Printing Office.

Commentaries

Biology education has come a long way since World War II, but it is time again to move beyond the status quo. Going beyond the status quo involves "recognizing the need to change; establishing a clear direction and purpose; and developing appropriate programs, policies and practices based on the new purpose." The next question inevitably is "How will this be accomplished?" particularly in the light of reduced funding.

Institutes geared to inquiry and problem solving, personal needs, and social issues could help (if and when funded) give new directions to biology teaching. Bybee's (1979b) suggestions for science education programs could be used as a springboard for school committees mandated to implement minimum competency standards.

A point missed in the paper is that the average biology teacher is so pressed for time in setting up labs, teaching large classes, and keeping abreast of the subject that there is not time left to be on top of everything from the availability of genetic counseling to the use of microcomputers.

Preservice teacher education should include all of the courses recommended, including those in the human relations aspects of biology teaching and the new areas outlined by the National Science Foundation. But where in the curriculum will they be squeezed? Inservice is a possibility, but inservice is a myth in many schools including mine. Teachers have always taken courses on their own at night or in the summer, in spite of the need of most to supplement their incomes. It might be possible to provide new, meaningful courses and encourage teachers to take them. A viable alternative might be to require a fifth year of college to provide these important aspects of teacher preparation before certification. It seems also that teachers should be expected to take and pass, initially and periodically, state board examinations to prove their knowledge in any subjects they teach.

Jane Abbott

How can I respond? I work in a school district of upper-middle class professionals. We are well funded and overequipped. Science education is still highly regarded in our community. We encourage the active participation of our students in individualized work through science projects. Many of our most successful student researchers are girls. We have very few minority or underprivileged students. Of a science faculty of eleven teachers, five are women, three of whom teach biology. Our girls have good role models in those teachers. We use a variety of instructional materials and curricula for all levels of instruction. Our students consistently score well on achievement tests. I do not think that we reflect many of the problems of the status quo outlined by Bybee and Kahle. What are its implications for me?

We are very traditional in our approach to science education. We stress content and concept. We reinforce both with adequate laboratory experiences. I might say we are even elitist. We are practically autonomous and do not interact in developing programs with other departments. We try to teach "the fundamentals." We are so involved in teaching content and concept that the implications of social and personal relevance in science teaching have passed us by. We strive to get our students into "good schools" and "good jobs." Where is there time for personal and social relevance if the competition for success becomes greater all the time? It seems as though we have so much to do with just basic instruction that personal and social relevance is just a distraction. We are caught up in the rush of living and are dragging our students along with us.

The ideas of Bybee and Kahle are so sweeping to be almost incomprehensible. I feel as though I either have to stop the train myself or jump off and let it pass me by while I get reoriented, reorganized, and reeducated. I think I cannot continue to teach what I think is important in the way I think it can be taught and yet evolve into a "personally and socially relevant" teacher within the meaning of this paper. Since I think many biology teachers today are caught in the same syndrome, how do we effect the changes envisioned here, both for ourselves and for the profession?

Roy K. Baer

Educators today have a greater challenge than ever—a challenge not only to educate, but to educate in a new and sometimes hostile environment. How will we meet these challenges? How will we expand the future of those entrusted to us? Of course there are many positive suggestions and proposed ideas of how we might better educate.

We in the classrooms must nod our heads with approval and enthusiasm to the ideas Bybee and Kahle present. However, when all points are considered, we must look at our own situations and build from them. Many of us are faced with a myriad of ideas on how to upgrade our present level of education, but we must do this with the reality that Johnny and Suzie just can't read. How then do we present a student-oriented program? Our hope lies in the educators charged with the great task of teaching a student to read. Once a program is established and supported that produces students who are willing and able to read, teachers of biology, and other subjects, will be able to implement the ideas proposed by Bybee and Kahle.

David P. Lapierre

I agree with Bybee and Kahle that going beyond the status quo in biology education requires that we first understand the present. Classroom teachers of biological science probably recognize most of the shortcomings outlined here, but do local boards of education, get the support needed to deal more effectively with some of these "status quo" problems.

A revival of National Science Foundation-supported academic year institutes and summer institutes to deal with going beyond the status quo would bring together curriculum development personnel and teachers to discuss and develop appropriate ideas and materials for this awesome, imperative task. I think that we need continued, expanded funding of the National Science Foundation to advance science in our society. I would like to be involved in going beyond the status quo.

James A. Meyer

I believe that this paper is the most important of all in this publication. It gives us a common purpose on which we can build programs, design policies, and direct practice. The emphasis on human ecology seems an appropriate path, but I urge that we stress individual involvement and recognition of how each person fits into the global ecology. Take your students outside the classroom and get them to look at their surroundings!

Also, we should make a concerted effort to incorporate into our programs those areas perceived as important in preservice training—decision-making and the concept of risk, nature of proof, values, preference, prejudice in thinking, and so on. If we do this, then facts and figures become the "icing" in our classes—while observing, questioning, and analyzing are the "cake."

How often I have heard from a student, "Science is not my subject," implying to me that "thinking" is not a part of his or her other classes. How can we help students to recognize problems and to identify and attack the elements of problems? To me, these are the "basics" of *any* class that a student might take. Additional universal basics are, of course, reading and writing. Poor reading and writing skills keep a student from understanding science. Must we help students attain these fundamental skills in our science classes? Yes, if *they* are to go beyond their own status quo.

It is important that science teachers be able to recognize their own weaknesses. How much minutiae do we ask students to learn each day? Is this our greatest weakness—that we give in to the easiest method of teaching, asking for the regurgitation of facts? Or do we struggle and work desperately hard to train student's minds? To me, the "discovery" or "inquiry" approach is the essence of science and the means by which students can carry themselves through a productive life. Proper teacher training is vital: teachers must understand the process of science before they can teach it. I also believe teacher training should combine both preservice general education and inservice professional duties so that theory can immediately be tested in practice. We must first ask ourselves to reach beyond the status quo before we ask our students to do the same.

Gordon E. Uno

ABOUT THE AUTHORS

ROGER W. BYBEE is Associate Professor of Education and Chairman of the Education Department at Carleton College, Northfield, Minnesota. He holds a Ph.D. in science education from New York University and B.A. and M.A. degrees from the University of Northern Colorado. Bybee has taught at the elementary, high school, and college levels. He has written three books on science teaching, including *Becoming a Secondary School Science Teacher* (with Leslie Trowbridge and Robert Sund). He is currently completing books on *Violence Values and Justice in the Schools* (Allyn and Bacon, 1981) and *Piaget for Educators* (Charles E. Merrill, 1982). His current research focuses on classroom discipline problems.

BENTLEY GLASS is Emeritus Distinguished Professor of Biology, State University of New York, Stony Brook, New York. He holds a Ph.D. from the University of Texas and A.B. and M.A. degrees from Baylor University. Dr. Glass was Professor of Biology at The Johns Hopkins University and Academic Vice President and Professor of Social Biology at the State University of New York, Stony Brook. Dr. Glass has been Editor of *The Quarterly Review of Biology* since 1958. He has served as President of the American Society of Human Genetics, Phi Beta Kappa, AAAS, NABT, and the American Association of University Professors. He was elected to the National Academy of Sciences in 1959. He is an honorary member of NABT.

FAITH M. HICKMAN is Director of the BSCS Center for Education in Human and Medical Genetics and Staff Associate for the BSCS, Boulder, Colorado. She holds a B.S. in Biology from West Virginia University and an M.A. in science education from the University of Colorado, Boulder. Ms. Hickman has served as Co-Chairperson of the Education Work Group, Commission for the Control of Huntington's Disease and Its Consequences, NIH; and Special Consultant on Education, American Academy of Pediatrics, Committee on Genetics and Committee on Health Education. She has been an Instructor of "Methods of Teaching Modern Biology" at the University of Colorado, Boulder and audiovisual review coordinator for *The American Biology Teacher*. She was NABT Director-at-Large in 1979 and Vice-President in 1980.

NEIL A. HOLTZMAN is Associate Professor of Pediatrics at The Johns Hopkins University School of Medicine, Baltimore, Maryland. He holds a Doctor of Medicine degree from the New York University College of Medicine and a B.A. from Swarthmore College, Swarthmore, Pennsylvania. Dr. Holtzman is Coordinator for Hereditary Disorders Services in Maryland and Director of the Johns Hopkins General Pediatrics Academic Development Program (supported by the Robert Wood Johnson Foundation). He serves on the Maryland Commission on Hereditary Disorders and on the Committee on Genetics of the American Academy of Pediatrics. He chairs the Social Issues Committee of the American Society of Human Genetics and sits on the Genetic Diseases Review and Advisory Committee on the Health Services Administration. His research interests include education, informed consent, neonatal screening, and alpha-fetoprotein screening.

PAUL DEHART HURD is Professor Emeritus of Education, Stanford University, Stanford, California. He holds an Ed.D. in science education from Stanford and an A.B. in botany and an M.A. in plant ecology from the University of Northern Colorado, Greeley. He has received honorary doctorates from the University of Northern Colorado, Ball State University, and Drake University. He has been awarded honorary life memberships in NABT, Phi Delta Kappa, NSTA, AERA, and other professional societies. He is a science education consultant to the National Science Foundation and a fellow of the Stanford-Northeast Asia-United States Forum on International Policy. He is a member of the Executive Committee of the BSCS and an advisor to the BSCS Center for Education in Human and Medical Genetics. He is currently conducting research on precollege science education in the People's Republic of China and on the status of junior high/middle school life science education in the United States.

JANE BUTLER KAHLE is Associate Professor in the Departments of Biological Science and Education and Director of Graduate Studies, Department of Education, Purdue University, West Lafayette, Indiana. She earned a B.A. at Wellesley College and an M.S. and Ph.D. at Purdue. She has taught biology at the secondary level for eight years and was a Senior Fulbright-Hays Lecturer to the Philippines in 1975. Dr. Kahle has served on many NSF and NIE projects and boards, including Project Synthesis, National Conference for Women and Minorities, and projects concerning instructional strategies for disadvantaged secondary school students. Her NABT activities and positions include Editorial Board, Publications Committee, Committee on Future Directions, and Director-at-Large. She was elected a Fellow in AAAS in 1981. She is the author of *Teaching Science in the Secondary School* (D. Van Nostrand, 1979), and *Biology Skills and Concepts* (Ginn & Co., forthcoming).

EDWARD J. KORMONDY is Provost and Professor of Biology, University of Southern Maine, Portland, Maine. He holds a B.S. from Tusculum College and M.S. and Ph.D. degrees from the University of Michigan. He was Instructor in Zoology and Curator of Insects, University of Michigan (1955-1957), Assistant to Professor of Biology, Oberlin College (1957-1968), Director of the Commission on Undergraduate Education in the Biological Sciences and Director, Office of Biological Education, American Institute of Biological Sciences (1968-1971); in addition to his faculty appointment in 1971, he served as Vice-President and Provost of The Evergreen State College from 1973-1978. He studied at the Center for Bioethics, Georgetown University in 1978-1979. He served as Senior Staff Associate, Directorate for Science Education, National Science Foundation in 1979. He is the author of some fifty research papers and reviews and ten books including *Readings in Ecology* (1966), *Concepts of Ecology* (1969, 1976), and *General Biology: The Integrity and Natural History of Organisms* (1977, with T. Sherman and others). He served as President of NABT in 1981.

JOSEPH D. McINERNEY is Associate Director of the Biological Sciences Curriculum Study (BSCS) and Staff Associate for BSCS Center for Education in Human and Medical Genetics. He holds a Master's degree in genetic counseling from the Graduate School of Allied Health Professions, Stony Brook University, and a B.S. in education from State University of New York, Cortland. Mr. McInerney taught special education in the public schools of Florida and New York for five years, while completing a Master's in education at Stony Brook. He has also served as a genetic counselor with the Long Island Sickle Cell Project and as an instructor at New York's Suffolk County Community College and at Stony Brook's School of Allied Health Professions. He is involved in several projects related to education in health and human and medical genetics, and is currently directing an analysis of ethical issues in the genetic counseling process under a joint grant from the National Science Foundation and the National Endowment for the Humanities.

GILBERT S. OMENN is currently Visiting Senior Fellow, Woodrow Wilson School of Public and International Affairs, Princeton University. At the time Dr. Omenn's paper was written, he was Associate Director for Human Resources and Social and Economic Services, Office of Science and Technology Policy, Executive Office of the President, Washington, D.C. Dr. Omenn earned an A.B. at Princeton University, an M.D. at Harvard Medical School, and a Ph.D. in genetics at the University of Washington, Seattle. Prior to his appointment with the Office of Science and Technology Policy, Dr. Omenn was Professor of Medicine at the University of Washington, Seattle, specializing in medical genetics. He did his internship and residency at Massachusetts General Hospital in Boston and was later a Research Associate at the National Institute of Arthritis and Metabolic Diseases. He was elected a member of the Institute of Medicine, National Academy of Sciences in 1978.

ELIZABETH J. ROBERTS is President of Television Audience Assessment, Inc., a nonprofit organization concerned with audience research and policy planning for cable and broadcast television. For the past six years she served as the Executive Director of the Project on Human Sexual Development, based at the Harvard University Graduate School of Education. She is former director of the Children's Television Unit of the Federal Communications Commission and served as director for the White House Conferences on Children and Youth. Her concerns include family life, the role and status of women, sexual learning, the influences of television on child development, and social and educational policy planning. Educated at Marquette University and Harvard University, Ms. Roberts has taught courses on sexuality and public policy, social and political philosophy, and ethics. Her publications include: *Childhood Sexual Learning: The Unwritten Curriculum; Family Life and Sexual Learning;* and a chapter in the updated Surgeon General's Report on *Television and Social Behavior.*